TRAGEDY AT PIKE RIVER MINE

TRAGEDY AT PIKE RIVER MINE

How and why 29 men died

REBECCA MACFIE

AWA PRESS

First edition published in 2013 by Awa Press, Unit 1, Level 3,
11 Vivian Street, Wellington 6011, New Zealand.

Copyright © Rebecca Macfie 2013

The right of Rebecca Macfie to be identified as the author of this work in terms of Section 96 of the Copyright Act 1994 is hereby asserted.

This book is sold subject to the condition that it shall not, by way of trade or otherwise, be lent, resold, hired out or otherwise circulated without the publisher's prior consent in any form of binding or cover other than that in which it is published and without a similar condition including this condition being imposed on the subsequent purchaser.

National Library of New Zealand Cataloguing-in-Publication Data
Macfie, Rebecca.
Tragedy at Pike River Mine : how and why 29 men died /
Rebecca Macfie.
Includes index.
ISBN 978-1-877551-90-1 (pbk.)—ISBN 978-1-877551-94-9 (epub)—
ISBN 978-1-877551-956 (mobi)
1.Pike River Mine Disaster, N.Z., 2010. 2. Coal mine accidents—New Zealand—Grey District. 3. Accident victims—New Zealand—Grey District. I. Title.
363.119622334099372—dc 23

Typesetting by Tina Delceg
This book is typeset in Minion Pro and National
Printed and bound in Australia by McPherson's Printing Group

Find more great books at awapress.com.

Produced with the assistance of

REBECCA MACFIE is a senior writer with *New Zealand Listener* and in a 25-year career in journalism has won over a dozen media awards. She has been senior writer and deputy editor of *The New Zealand Herald* publication *The Business*, is a former editor of *Unlimited*, a former staff writer for *National Business Review*, and has written extensively for *The Independent Business Weekly*, *North & South* and *Safeguard*. She is a resident of Christchurch; several of her articles on that city's major earthquakes were published as a BWB Text in April 2013.

For the Pike 29

Contents

	Author's Note	1
	Pike People	6
	Prologue	11
ONE	Pike Dream	23
TWO	Great Expectations	37
THREE	Early Warnings	47
FOUR	Trouble from the Start	65
FIVE	Management Blues	83
SIX	Many Whistles Blowing	97
SEVEN	Too Big to Fail	117
EIGHT	Marching to Calamity	137
NINE	Who Will Say Stop	167
TEN	November 19	183
ELEVEN	Five Days	201
TWELVE	Entombed	223
	Timeline	245
	Glossary	249
	Endnotes	253
	Index	267

Author's Note

I imagine most New Zealanders are able to recall the moment when they heard that a large group of workers was trapped inside Pike River coal mine following an explosion. I was at Wellington Airport waiting for a flight home to Christchurch, and happened to glance up at a television screen as the sketchy news was relayed.

I knew only a little of Pike – as a journalist I had not written articles about it, but I had tramped in the Paparoas and knew roughly where the mine was, and my father-in-law owned shares in the company and would often write to my husband and me of his belief that it would be a star performer. I was aware from news headlines that there had been repeated delays and cost blowouts in the mine's development, and that the company seemed to be forever going to its shareholders to ask for more capital.

Throughout that Friday night we tuned in repeatedly for news updates, in the hopeful but naïve expectation that the men – exact numbers still unknown – would come out alive. I'd read about China's appalling record of underground coal mine fatalities, and knew that the names Kaitangata, Brunner and Strongman were shorthand for some of the worst mining tragedies in New Zealand's history. But Pike was a modern company, and the date was November 19, 2010. Surely this could not be a repeat of such historic catastrophes?

Over the weekend my editor at *New Zealand Listener*, Pamela Stirling, senior writer Joanne Black, photographer David White and I planned how we would cover the story. As a weekly magazine, we weren't in the business of reporting blow-by-blow updates. We were due to go to print on the Wednesday night, and our job as a current affairs weekly was to delve into the background and begin digging out answers to the key question: how can this have happened? It was decided that Joanne and David would travel to the Coast and gather information and images from the region. I would stay in Christchurch and begin building up a picture of the company, based on corporate documents and the insights of people with knowledge of the mine.

By the time we went to print that week, the mine had exploded a second time and it was known that 29 men were dead. It was also clear that Pike had understated its most critical risk – methane gas – and had repeatedly failed to meet its own grandiose forecasts. Far from being the showcase of modern mining it had branded itself as, the project had lurched from one major setback to the next throughout its short history. It seemed that virtually everything that could have gone wrong in the development of the mine had gone wrong; and then, with coal extraction having barely started, it exploded. I formed the view that the disaster was not an 'accident', but rather a corporate failure of the worst order.

The Pike story filled me with rage and grief. As a mother of two – at the time, my son was just two years older than the youngest of the Pike 29 – it seemed to me a parent's unimaginable horror. As a journalist, it was one of the most important stories of my generation. As a New Zealander, it was a source of shame.

The tragedy of Pike River mine will be the subject of historical investigation and academic study for decades to come. It has already been the subject of a Royal Commission of Inquiry report, and two court decisions. This book has been written with the aim of making the Pike story understandable to a wide audience – for the families, friends and colleagues who grieve for the 29 men, for New Zealanders who want to understand how such an avoidable disaster occurred in their midst in the 21st century, and for the leaders of businesses and organisations who must learn from it.

Writing this book has been the most difficult task I have ever undertaken as a journalist, but I have been blessed with the support of a large number of people. In particular, Gerry Morris has been a vital participant in the project. Gerry is a loyal son of the West Coast, and three members of his family worked at Pike. He shared my determination that the pain of Pike must not be forgotten. He assisted with research, introductions and insights into the culture and history of the Coast; he vouched for me in a community still raw with sorrow; he shared his contacts and dug out information that I would have missed; he gave me instant feedback on draft chapters. I doubt I could have done it without his constant help, counsel and encouragement.

I am also deeply indebted to Pamela Stirling, who supported my pursuit of the Pike story through the *Listener*. She backed my decision to write this book, enabled me to take leave from my job to focus on it exclusively, and agreed to my repeated pleas for more time.

I am enormously grateful to Harry Bell, who spent many hours patiently instructing me on underground coal mining, reading my draft chapters and saving me from many mistakes. I feel blessed to have had the benefit of Harry's kindness, experience and wisdom.

To Therese and Denis Smith, whom I came to think of as my West Coast family, my heartfelt thanks. Whenever I was on research missions, Therese and Den would envelop me in the love and warmth of their Greymouth home and care for me as one of their own.

Thanks also to the Bruce Jesson Foundation for granting me their senior journalism award in support of my work on this book.

In the course of my work for the *Listener* I spent many weeks in 2011 and 2012 reporting on the proceedings of the Royal Commission on the Pike River Coal Mine Tragedy. This book has benefited from the knowledge and documents I obtained as a result. Anna Hughes, the media officer for the commission, worked hard to facilitate extensive and detailed reporting of the commission, and I greatly appreciated her efforts.

I have also drawn on a large number of documents available in the public domain, as well as others I obtained through the Official Information

Act or that came through confidential sources. I conducted interviews and email correspondence with many people associated in various ways with the Pike story. It was gratifying that they were willing to share their knowledge and insights with me despite, in many cases, having already gone through lengthy interviews with Department of Labour and police investigators, and given evidence to the commission.

My thanks (in no particular order) to: Trevor Watts, Robin Hughes, Rob Smith, Troy Stewart, Dave Stewart, Bernie Lambley, Quintin Rawiri, Tony Kokshoorn, Matt Coll, John Robinson, Jane Newman, Murry Cave, Barry McIntosh, Dene Murphy, Scott Campbell, Peter Sattler, Russell Smith, Dan Duggan, Denis Smith, Rick Durbridge, Kevan Curtis, Brent Mackinnon, Nan Dixon, Anna Osborne, Sheryl and Johnny Griffin, Gordon and Ian Dixon, Annette Hughson, Robyn Hennah, Bernie Monk, Kath, Olivia and Alan Monk, Carol Rose, Malcolm and Jane Campbell, Nick Davidson, Lynne and Trevor Sims, Hugh Logan, Chris Roberts, Tony Frankham, David Salisbury, Chris Stone, Paul Caffyn, Peter Gunn, Brian Roulston, Roger O'Brien, Graham Mulligan, Graeme Duncan, Peter Liddle, Dave Bennett, Terry Bates, Richard Cotton, Evan Giles, Joe Edwards, Seth Tiddy, Kobus Louw, Les McCracken, Masaoki Nishioka, Nigel Slonker, Kevin Cash, Dave Cross, Sean Judd, Mark Watson, John Canning, Gary Knowles, Peter Read, Catriona Bayliss, Denise Weir, Shane Bocock, Phil Smith, Stuart Nattrass, John Dow, Kevin Brown, Jonny McNee, Don McFarlane, Les Tredinnick, Greg Borichevsky, Andrew Knight, Andrew Harvey-Green, John Key, Darren Brady, Dave Cliff, John Fisk, Stewart Bell, Keith Teasdale, Peter Haddock, Paddy Blanchfield, Chris Yeats, Rose Green, Joy Baker, Brian Small, Nick Thompson and Mark Pizey. I am grateful also to Lauryn Marden, Tara Kennedy and Kim Joynson for consenting to my use of extracts from their statements to the Royal Commission. I have also had invaluable help from many confidential sources. They can't be named for obvious reasons but they will know who they are – thank you for your information and guidance.

Bernie Lambley (Terrace Mine) and Andrew Holley (Roa Mine) generously agreed to guide me through their operations to help build my understanding of underground coal mining. I'm grateful also to Colin Smith for his support and liaison, David Butler for his advice, and Stacey Shortall for her cooperation. Others who helped directly or indirectly include Bryn Somerville, Matt Curtis, George Colligan and Catherine Milne.

A number of people went out of their way to make photographic and diagrammatic material available to help my understanding, and also for publication in this book, including Stewart Nimmo, Matt Coll, Trevor Watts, Martin Hunter, Tony Kokshoorn and Patrick McBride.

I was unable to speak to a number of people who are central to the Pike story. Tony Radford refused to be interviewed, as did Ray Meyer, Arun Jagatramka and Stephen Rawson. Dipak Agarwalla ignored my request for an interview. Peter Whittall declined my request on legal advice: he is facing charges under the Health and Safety in Employment Act, which are to be heard in 2014. Gordon Ward did not respond to requests put to him on my behalf. Doug White refused to be interviewed.

My thanks to Mary Varnham, Sarah Bennett, Kylie Sutcliffe, Lee Slater, Ruth Beran and Angela Radford – it has been a privilege to work with the team at Awa Press. Mary recognised very soon after the Pike disaster that this was a story of enduring significance. I am honoured that she asked me to write it, and grateful that she persisted when earthquakes and other disruptions got in the way.

I could not have completed this book without the family members and friends who have backed me and understood the importance and gravity of the topic. My thanks especially to Grace, who spent hours proofreading chapters; to Charles for his support and interest; to Bob, Helen and Belinda, who gave me feedback on drafts; and to Alex, who read early chapters and supported Gerry's passion for the Pike story.

Most of all, my loving thanks to Neil, who has had to put up with my frequent absences and single-minded focus on Pike, cared for me through what has been an intensive period of work, and believed in my ability to produce a book that would help bring insight into one of the most grievous events in New Zealand's history.

Rebecca Macfie

Pike People

A large number of people were influential in the planning, conception and operation of the Pike River coal mine. To assist the reader, the following is a brief guide to those who played a significant role in the Pike story as directors, senior managers, representatives of major investors, key consultants and regulators. The names appear in alphabetical order, not in order of significance.

Dipak Agarwalla Director of Pike River Coal Ltd from September 2005 until September 2011. Agarwalla represented Saurashtra Fuels, a significant investor in Pike, on the board of directors.

Terry Bates Geologist and exploration manager of Otter Minerals Exploration Ltd, a subsidiary of Mineral Resources New Zealand Ltd, a company chaired by New Zealand-born accountant Tony Radford. Bates conducted exploration of the Pike coalfield in the early 1980s, but had no further involvement.

Behre Dolbear Australia Pty Ltd Mining consultancy that produced technical reports on the Pike River mine proposal for the guidance of prospective shareholders in the 2007 initial public offering of mine company shares. It also reviewed the mine operations for major shareholder New Zealand Oil & Gas Ltd in May 2010.

John Dow Director of Pike River Coal Ltd from February 2007 and chair of the board from May 2007 until September 2011; a retired gold mining company executive.

Graeme Duncan Australian mining consultant whose former company Minarco (previously called AMC) produced a final feasibility study for Pike River Coal Ltd in 2000, which was paid for in shares. Duncan was a member of the Pike River Coal Ltd board of directors from 1999 until December 2006. Minarco also provided other technical advice to Pike, including a 2005 update of the original feasibility study.

Steve Ellis Pike River Coal production manager from the beginning of October 2010. Ellis also informally carried out the duties of statutory mine manager for several weeks before the explosion of November 19. He was formally appointed statutory mine manager by the receivers of Pike River Coal Ltd in May 2011.

Michael Firmin Department of Labour health and safety inspector with regulatory oversight of Pike River coal mine from 2005 (before mine development had begun) until mid 2008.

Tony Goodwin Pike River Coal engineering manager from 2005 until March 2009.

Nick Gribble Pike River Coal engineering manager from April 2009 until August 2010.

Peter Gunn Geologist and mining consultant who was heavily involved with the development of plans for a coal mine at Pike from the mid 1990s, and continued to provide advice until 2006.

Arun Jagatramka Pike River Coal Ltd director from July 2007 until September 2011, representing major shareholder Gujarat NRE, an Indian coke making and coal mining company.

Dick Knapp Pike River Coal human resources manager from January 2009.

Bernie Lambley Pike River Coal acting production manager from June 2010 until September 2010.

Mick Lerch Pike River Coal statutory mine manager from January 2010 until June 2010.

Ivan Liddell Pike River Coal environmental manager from February 2005 until January 2011.

Kobus Louw Pike River Coal tunnel and production manager from May 2007 until October 2008, and then statutory mine manager until February 2009.

Sanjay Loyalka Pike River Coal Ltd director from November 2009 until September 2011. Acted as an alternate director for **Arun Jagatramka**.

McConnell Dowell Major construction company that was Pike River Coal's largest subcontractor, building the 2.3-kilometre stone tunnel and 111-metre ventilation shaft, undertaking remedial work to fix the shaft after it partially collapsed, and carrying out other underground contracting services.

Les McCracken Mining consultant who acted as Pike River Coal project manager on the construction of the road and tunnel/shaft contracts from 2004 until 2007.

Ray Meyer Pike River Coal Ltd director from June 2000 until December 2010.

Stuart Nattrass Pike River Coal Ltd director from February 2007 until September 2011.

Kevin Poynter Department of Labour health and safety inspector with oversight of Pike River mine from July 2008.

Tony Radford Chair of Pike River Coal Ltd's board of directors until April 2006, and a director until June 2011. Also chair of New Zealand Oil & Gas Ltd, the company that promoted the Pike River mine project and remained the largest shareholder, with 31 percent, after the initial public offering of Pike shares in June 2007. Radford remained chair of New Zealand Oil & Gas Ltd until 2012.

Stephen Rawson Pike River Coal Ltd director from November 2000 until June 2006.

Udo Renk Pike River Coal technical services manager from January 2007 until May 2008.

Robb Ridl Pike River Coal engineering manager from August 2010.

Neville Rockhouse Pike River Coal safety and training manager from December 2006 until April 2011.

David Salisbury Managing director of New Zealand Oil & Gas Ltd from April 2007 until December 2011.

Nigel Slonker Pike River Coal statutory mine manager from April 2009 until September 2009.

URS New Zealand Geotechnical consultancy that advised both Pike River Coal and McConnell Dowell.

Valley Longwall International Australian drilling company contracted by Pike River Coal to conduct in-seam drilling to assist with understanding of the coal seam.

Pieter van Rooyen Pike River Coal technical services manager from February 2009 until early November 2010.

Gordon Ward New Zealand Oil & Gas accountant from the late 1980s and finance manager from the late 1990s. Ward was the key driver of the project to develop the Pike River mine, overseeing planning and regulatory issues from 1998. He became a director of Pike River Coal Ltd in July 2006 and chief executive in January 2007. He continued in both positions until September 2010, when he was dismissed by the Pike River Coal Ltd board.

Doug White Pike River Coal operations manager from January 2010 until October 2010, when he was promoted to general manager; also carried the role of statutory mine manager from June 2010 until May 2011.

Peter Whittall Pike River Coal's first employee, hired in 2005 as general manager, mines. Whittall oversaw design and development of the mine as a start-up operation. Also filled the role of statutory mine manager from September 2009 until December 2009. He became chief executive of Pike River Coal Ltd in October 2010. He finished working for the company (in receivership) in November 2011.

Prologue

It came with a powerful blast of pressure and a flash of white light that lit up the tunnel. The surge of debris-laden air blew Russell Smith's helmet off his head and smashed him in the face with a concussive blow. Instinctively, he sought refuge by ducking down behind the metal door of the loader he had driven in from the tunnel entrance a few moments earlier.

Smith had been heading underground to join his 'C crew' workmates at the coalface in the north-west part of Pike River mine. Earlier in the afternoon he had been at home in Cobden with his wife Jo, and had forgotten that, because it was a Friday, his shift started an hour earlier than normal. Realising he was late, he had raced up the Grey Valley to the mine site, where he bumped into the acting underviewer, Conrad Adams, who was still on the surface. Adams told him to drive a loader up to the face; he then headed underground, saying to Smith, 'I'll meet you up there.'

Smith was further delayed by trouble with the loader. There was a problem with the accelerator, and then he struggled to get the bucket on.

He was about 1.5 kilometres in when the explosion bore down on him. As he crouched for protection in the loader, rocks blasted at the machine and ripped across his back and neck. There was neither heat nor flames. He thought a compressed air line must have blown, but then he started battling for breath. He knew he needed to use his self-rescuer – the emergency kit that all miners carry strapped to their hips, providing 30 minutes of oxygen – but couldn't get at it because of his cramped position. He got out of the vehicle, tried to straighten up and release the breathing apparatus, and then collapsed to the hard tunnel floor.[1]

The explosion hit Daniel Rockhouse as a hard shove against the chest, forcing him backwards on to the ground. The sound was deafening. At first he thought his loader had blown up. He had just driven it from where he and the rest of C crew had been working down to the utilities area known as 'pit bottom in stone'. He had been sent by his supervisor, mine deputy Dan Herk, to fetch some gravel with which to repair a rough section of roadway; once that was done, the crew would be able to re-position machinery to enable coal to be cut on the following Monday morning.

On the way down he'd come across Riki Keane at Spaghetti Junction – a bottlenecked place of pipes and electric cables. Keane was trying to start a broken-down loader, and Rockhouse had shot back towards the mine workings to get some oil to help restart the vehicle.

Rockhouse paused to chat with Conrad Adams, who was heading up into the mine in a driftrunner.

'Sweet as,' Adams said, after Rockhouse explained that he'd been despatched to get gravel. 'I'll see you when you get back.'

'Righto,' Rockhouse replied, and continued down the tunnel.

He had just parked the loader in a fuel bay and begun to top it up with diesel and water when he was thrown backwards, hitting his head as he fell. Staggering back to his feet, he saw that the loader was intact and the motor was still running. Within seconds, thick pungent smoke came from around the corner and engulfed him. There was a strong smell of diesel and he was certain he was breathing in carbon monoxide, the product of a fire somewhere in the mine. It was disgusting to swallow, and his eyes and nose ran. He put on his self-rescuer but couldn't draw enough air from it. He was panicking, angry with the bloody thing. He ripped it out of his mouth, and then fell to the ground.

His body became tingly and felt as if it were shutting down; he became numb from the neck down and was terrified. He thought he would die. 'Please, don't do this,' he pleaded into the smoke-filled blackness. It was deathly quiet. 'Is there anyone out there? Help. Help. Help.' No one answered.

He lost consciousness. It was perhaps 50 minutes before he revived as air was drawn back into the mine from the portal, almost two kilometres away. He managed to stand up briefly, then fell again. 'Get up, you bastard,' he screamed at himself. 'Get up, get up.'

Somehow he managed to propel himself to the side of the tunnel, where there were pipes carrying compressed air and water. He turned on an air valve, which released a sluggish flow of oxygen; normally the pressure would be enough to blow out a man's eyeballs. He bathed his face and lungs in the stuff, and knew he had to get out.

Through smoke so thick he could see only his hand in front of his face, he found his way to an underground phone and rang the mine emergency number, 555. It went to an answering machine. He screamed at the phone and tried another number – 410, the number for Pike's surface control room – and reached controller Dan Duggan and Pike's mine manager Doug White. There had been an explosion, he told them. 'Help me.'

White, a Scotsman with years of experience in Australian coal mines, reassured him: 'You can make it. Get out, stay low and get to the fresh air base and make contact there.'

He continued groping down the tunnel in the darkness, turning on air valves as he went in the hope that might dilute the gas. His younger brother Ben was underground – they had spoken earlier in the day – but he knew going back up into the mine to find him would be suicide.

About 300 metres on he found Russell Smith on all fours. Rockhouse grabbed him by the hair and lifted his head back to see who it was, and saw Smith's eyes rolling to the back of his head. He tried to put Smith's self-rescuer into the man's mouth, but he was so weak it fell out. He grabbed Smith from behind and dragged him to the shipping container that served as a fresh air base. But there was no refuge of any kind there:

there was neither a working phone nor a supply of self-rescuers, and the compressed air valve didn't work. The place had been decommissioned.

Enraged, he returned to Smith, whom he had propped up against the side of the tunnel. 'Can you walk?' he asked. Smith could only groan. 'Screw this. We're getting out of here,' Rockhouse said, and hauled the other man to his feet. Together they staggered towards daylight and fresh air, with Rockhouse propping up Smith and using his free hand to grasp the rail of the conveyor for support. Rockhouse frequently looked back up the 2.3-kilometre tunnel towards the mine workings in the hope of seeing the headlamps of other men following, but there were no lights to be seen.[2]

It was 5.26 p.m. when they reached the portal, an hour and 42 minutes after the explosion. To Rockhouse their escape had seemed to take forever. No one was there to meet them when they emerged with the stink and the red, bulging-eyed look of those who have suffered carbon monoxide poisoning. There was just the soft rustling of the dense bush that cloaked the Pike coalfield and the surrounding mountain range, and the rush of White Knight Stream. Rockhouse called for help from a phone at the portal.

Pike contractor Shane Bocock arrived moments later. With him were two of his employees, Dylan Hutton and Steve Wakeford, and Rockhouse's father Neville, the mine's safety and training manager. Daniel Rockhouse was stooped and exhausted; Russell Smith was incoherent. They were bundled into Bocock's utility vehicle and ferried to a waiting ambulance.[3]

Dan Duggan hadn't noticed the noise on the phone at the time. From the surface control room, part of a compound of discrete low-rise mine administration buildings set amid ancient rainforest, he had made a call just after 3.44 p.m. on the underground intercom system with the intention of telling the miners he had turned a water pump back on. The pump had been off because of a planned maintenance shutdown since 12.20; with water back on, the men could start mining again.

The DAC – digital access carrier – system transmitted his voice to 15 or more points in the mine, and any number of workers underground might have responded to his call. It had been installed only a couple of months earlier, and was proving an excellent means of maintaining reliable communication between the control room and the mine.

It was Malcolm Campbell, a Scottish engineer working with a continuous mining machine at the far west of the Pike development, who answered. But their conversation was cut short.

'Hello, Dan. Who you looking for?' Campbell said.

'Just after the ABM and roadheader,' replied Duggan.

The ABM – Alpine Bolter Miner – machine was being manned by C crew, which included his brother Chris Duggan. The roadheader crew of Blair Sims and David Hoggart were on maintenance tasks in a location just to the south of the ABM crew.

Above the static and noise of alarms going off around him, Duggan didn't hear the growling boom down the line. After his initial greeting, Campbell had stopped responding.

'Hello sparkies,' Duggan called. No one answered. 'Hello underground, any sparkies?' A minute of noisy, raspy silence followed.

'Hello monitor place,' he called, hoping to hear from Allan Dixon, Peter O'Neill or Keith Valli, the three men working that afternoon on Pike's recently commissioned hydro monitor, a specialised mining system that carved coal from the face with a laser-like jet of water.

Nothing. 'No cunts ringing,' Duggan cussed.

'Hello, anyone underground?'

'Hello. Monitor place, anyone underground? Anyone?'[4]

The alarms in the control room told Duggan that the power was down, and that gas, pump and ventilation monitoring had been lost underground. He had a bad feeling: it was odd that no one was responding. Calls over the DAC system were almost always answered promptly, and Duggan knew there was a back-up battery to keep the system operational even if the power was off. He informed White of the loss of communications and power. A few minutes later he asked whether he should put the

Mines Rescue Service on standby; it was based near the small coastal settlement of Rapahoe about 45 kilometres away. He still couldn't raise a response from anyone underground, and by then another worker had mentioned a strange smell in the air.

It was not yet necessary to contact the rescue service, White told him. Someone would be sent to the mine portal, about a kilometre up the narrow winding road from the administration building, to check things out.[5] Duggan respected his boss's judgement; he held back from making the call.

White, with still no idea that a catastrophe had occurred underground, went to his office and wrote some emails. One, to an Australian coal industry recruitment executive Garry McCure, said simply: 'Free now. Doug.' White had earlier written to McCure about his loss of trust in Pike's chief executive, Peter Whittall. Although he had been at Pike for only ten months, he'd told McCure he wanted out. His short email at 4.03 p.m. was in response to an earlier request from McCure to let him know when he was free to speak.[6]

It was 4.11 p.m. when Pike electrician Mattheus Strydom was sent into the mine to investigate why power and communications had been lost. Nothing had been heard from underground since the loss of contact with Malcolm Campbell at 3.44. No one on the surface knew that Rockhouse and Smith were, at that time, still lying unconscious, or what state the other 30 or so men were in.

Strydom was nervous. He paused before driving his vehicle into the portal. Things did not feel right. From his 26 years as a coal miner in his native South Africa he knew of the horror of explosions caused by the methane gas that is emitted by coal when it is mined. Pike was a gassy mine; he feared this wasn't just an ordinary power outage, that something major had happened underground.

There was a smell of cordite, and as he proceeded up the tunnel the air felt thick. When he had driven in about 1.5 kilometres he saw the light of a vehicle in the distance, and as he continued on he saw the motionless body of a man. By then Strydom's machine had started faltering and he was struggling for breath; he hadn't taken a self-rescuer with him. He thought the man in front of him was dead and that he, too, would die. He reversed as fast as he could down the tunnel until he reached a place where he could turn the vehicle around. When he reached the portal he rang the control room. 'You better call Mines Rescue,' he said. 'We've had an explosion and I've seen a man lying on his back in the road.'[7]

White now accepted he had a major event on his hands, and at 4.26 – after 41 minutes in which there had been no power, communications or response from anyone underground – Duggan was finally able to call the Mines Rescue Service. The phone number was normally written in bold type on a piece of paper on the control room wall. Unbelievably, Duggan found it was missing, and so with shaking hands he had to look it up in the phone book. He told himself, 'Get your shit together,' and made the call.

A few minutes later he phoned 111, remaining calm as the minutes ticked by while the emergency operator looked up the location of the mine on the map, double-checked the phone number and address, and queried him on why he thought there had been an explosion. 'We just don't know how bad this is,' Duggan told the operator. 'It could be the worst, you know … It's not gonna be good, I'll tell you that.'[8]

At 4.40 Duggan and White received the first and only contact from underground since communication had been lost almost an hour before. It was the call from Daniel Rockhouse, pleading for help.

Nan Dixon came home from visiting friends across the creek at about 4.45 that Friday afternoon. It was time to get tea on for her son, Allan; she was expecting him home from his shift at Pike at about seven. He'd been working at the mine for about two and a half years, since early on in the project's development. Four days a week he lived with her in the small, immaculately kept weatherboard house in which he and his three siblings had grown up, in the mining township of Rūnanga. On his days off he would head home to be with his partner Robyn Hennah in Upper Moutere, near Nelson.

Nan had lived all her 79 years in Rūnanga, a small town cradled between the jagged spine of the Twelve Apostles and the folding, faulted expanse of the Paparoa Range. The place was still surrounded by breathtaking natural beauty, but the years had been hard on it. Once, the area's big state-owned coal mines had provided work for hundreds of local men, with Rūnanga's commercial centre servicing the bulk of their daily needs; there had been a baker, a grocer, a drapery, a fruit and vegetable shop, a library, a workingmen's club, and a thriving miners' cooperative store. By 2010, even with the development of the new Pike River mine and the nearby government-owned Spring Creek coal mine, also underground, Rūnanga was a scruffy shadow of the vibrant community in which Nan had grown up, married and raised her four children. There were only two shops, and the primary school roll had shrunk from close to 300 children to under a hundred.

Like her father Walter Bell and her brother Harry, Nan's husband Jack had worked at the old Liverpool underground mine at Rewanui, but he had died of stomach cancer at 52, leaving her as a young widow and with two teenage children still living at home. Allan had also worked at Liverpool, Strongman and other underground mines until he left the industry in the mid 1990s. When in 2008 he decided to come back to the West Coast, work at Pike and live part-time with her, it was almost like having Jack back after almost 30 years alone. She was a superb baker of sweet treats and Allan's lunchbox was an object of envy among the men underground.

It was always a relief to see her son come in at the end of a shift, as it had been to see Jack. The region wore the grim scars and memorials of underground mining catastrophes: 65 dead in the Brunner explosion in 1896; nine dead in the Dobson mine in 1926; five lost in an explosion at Kaye's mine in 1940; 19 lost at Strongman in 1967; four in the Boatmans mine in 1985; three killed in two incidents at Mt Davy in 1998; two dead in accidents at Roa and Black Reef mines in 2006.[9] Allan himself had once been buried by rock in a mining accident.

Like everyone middle-aged or older on the West Coast, Nan had vivid recall of the sound of police cars and grief the day Strongman blew up. In her street alone, three men had been killed in the disaster.

'You always worried. If you hear of any accident, or if you see police cars and ambulances screaming along, a mining accident is the first thing that comes into your mind,' she would say.

Lately she'd been especially worried about Allan's safety at Pike. Earlier in the year he had suffered a knee injury in the mine that had required surgery, and she felt he'd been chased back to work before it had time to properly heal. And over the previous two or three months he'd been coming home from work looking terrible and feeling too sick to eat, often going straight to bed with a Panadol. He thought he was suffering from the effects of diesel fumes underground.

'I used to say, "Get out before it's too late." He said, "Mum, they tell you to shut your face and get the coal out. They just don't want to listen."' He'd often talk on the phone to his brother Gordon, who lived in Christchurch, about the frustrations underground, with machinery that didn't work and methane levels that rose quickly. He told Gordon it was a waste of time putting in complaints about safety issues at Pike because they were ignored.

Allan was 59, and was planning to work at Pike until Christmas. He was there for one purpose only – to pay off the debts he was carrying. The good wages at the mine gave him his best shot at achieving that goal.

Nan had not long started preparing Friday's evening meal when the phone rang. Her daughter Annette Hughson was calling from Rangiora, not far from Christchurch, on the other side of the South Island. She had heard from Gordon there was news on the internet of an explosion at Pike. Nan felt as if she would collapse: she knew Allan would still be underground.[10]

Annette also called Nan's best friend Nellie O'Neill to ask her to sit with her mother. She hadn't realised that Nellie's son Peter was underground too.

Word quickly leaked out into the Grey Valley, and up and down the skinny green margin of habitable land between the mountains and the Tasman Sea. It reached the kitchens and meeting places of Greymouth, Ngāhere, Blackball, Dunollie, Barrytown, Paroa and Hokitika. It seeped out to the far-flung families of Pike men in Southland, Christchurch, Australia, Scotland and South Africa. Some were alerted to disaster by the sound of sirens and helicopters; others were phoned by friends or family who had heard about the explosion on the news.

Anna Osborne, who could look directly across the Grey Valley to the Paparoas from her weatherboard Ngāhere home, was having a raspberry and coke with her 15-year-old son Robin at the local pub when she heard. Robin had just started a job on a local dairy farm and was shouting her a celebratory drink. He was looking forward to buying his father Milton – a burly teddy bear of a man who was deputy fire chief of the Ngāhere volunteer fire brigade – a beer when he knocked off work that day.

Fifty-four-year-old Milton had set up his own subcontracting company with the support of a local businessman and fellow Grey District councillor, Peter Haddock, a few months earlier. He was engaged by Pike to lay the fluming pipes that would carry the slurry of coal and water down the mine. Over the previous couple of weeks the workload had been enormous – 16-hour days, usually – and he had been coming home looking grey and exhausted, eating his meal and then falling asleep in the bath. The pressure was on to get the mine into production and start getting coal out. But the money was good, and Osborne felt positive that his move to self-employment would allow his family to build some financial security. If he had any worries about safety he didn't reveal them to Anna, who he knew was anxious about him being underground.

Anna and Robin were having a round of pool when Peter Haddock rang Anna's cell phone to ask if Milton were home from work. No, Anna said, but he wouldn't be far away.

'Haven't you heard?' Haddock said.

Electrified with shock, Anna took off to Pike, passing Milton's ute, which was parked where he'd left it that morning, and reaching the mine gate at 5.05 p.m. She was allowed through, along with Lynne and Trevor Sims, the parents of miner and talented local sportsman Blair Sims; the Sims had also raced up the Grey Valley in search of information. These were the only family members let on to the site before cordons were set up.

Once there, Anna Osborne refused to leave until her husband came out of the mine. She camped in the workers' smoko room near the miners' bathhouse for four nights, sleeping in snatches on a table, waiting and hoping for Milton to walk down the hill. The wives of Pike workers baked and sent supplies to sustain her through her vigil.[11]

But Milton was almost certainly dead, along with the 28 others who were never seen by their families again. They had been killed either by the force of an explosion fuelled by an enormous accumulation of methane, or by the aftermath of noxious gases and oxygen depletion.[12]

Only Daniel Rockhouse and Russell Smith survived. Five days after the explosion of November 19, 2010, the mine blew up a second time, and then again and again on November 26 and November 28. The remains of the 29 men have never been recovered.

Conrad Adams, 43, of Greymouth
Malcolm Campbell, 25, of St Andrews, Scotland
Glenn Cruse, 35, of Greymouth
Allan Dixon, 59, of Rūnanga
Zen Drew (Verhoeven), 21, of Greymouth
Chris Duggan, 31, of Dunollie
Joseph Dunbar, 17, of Christchurch
John Hale, 45, of Hokitika
Daniel (Dan) Herk, 36, of Rūnanga
David (Dave) Hoggart, 33, of Greymouth
Richard (Rolls) Holling, 41, of Blackball
Andrew (Huck) Hurren, 32, of Hokitika
Koos Jonker, 47, of Limpopo, South Africa
Willie Joynson, 49, of Maryborough, Queensland, Australia
Riki (Rik) Keane, 28, of Greymouth

Terry Kitchin, 41, of Rūnanga
Samuel (Sam) Mackie, 26, of Christchurch
Francis Marden, 41, of Barrytown
Michael Monk, 23, of Greymouth
Stuart (Stu) Mudge, 31, of Rūnanga
Kane Nieper, 33, of Greymouth
Peter O'Neill, 55, of Rūnanga
Milton (Milt) Osborne, 54, of Ngāhere
Brendon Palmer, 27, of Greymouth
Benjamin (Ben) Rockhouse, 21, of Singleton, New South Wales, Australia
Peter (Pete) Rodger, 40, of Perth, Scotland
Blair Sims, 28, of Greymouth
Joshua (Josh) Ufer, 25, of Charters Towers, Queensland, Australia
Keith Valli, 62, of Nightcaps, Southland.

The disaster at Pike River coal mine was New Zealand's worst coal-mining tragedy since 43 were killed at Ralph's Mine near Huntly in 1914.

This is the story of Pike, and how and why 29 men came to die there.

ONE
Pike Dream

The right to explore the Pike River coalfield had sat for years gathering dust within the recesses of Tony Radford's complex corporate web. Nothing was going to happen unless a partner could be found with deep pockets and the reservoir of expertise and infrastructure needed to turn the thick seam underlying the rugged and precipitous Paparoa Range into a working coal mine.

No one doubted the potential importance of Pike's coal. It was an enormous untapped resource, formed from a thick mire of peat inundated by seawater, buried deeply and heated for nearly 45 million years, and then thrust upwards by the seismic forces that heave and twist at the mountainous spine of the South Island. Spanning between a high sheer escarpment to the west and the Hāwera fault to the east, the seam lay 13 metres thick in places, ran six and a half kilometres from north to south, and was up to two kilometres wide.

This was no ordinary coal. It was 'beautiful'[1] – a bright, lustrous coking coal with properties that would be sought after by international steelmakers, who rely on coke as a source of fuel for their blast furnaces, and of carbon to bond with iron to produce steel.

Pike's low ash and high fluidity[2] made it a unique and potentially highly valuable resource – if a viable method could be found to extract it and transport it from deep in the West Coast mountains to far-flung world markets.

But Radford was no miner, and the increasingly convoluted network of companies he had built – initially in partnership with geologist Dave Kennedy – had no expertise in the highly specialised and technically challenging business of coal mining. Diminutive and socially reserved, Radford could scarcely be more removed from the raw and unpretentious West Coast townships that had prospered and declined with the fortunes and tragedies of the mining and logging industries. He was born in Palmerston North and had studied accountancy before moving to Australia and carving out a small empire of mineral exploration interests. His developing obsession was the formation of a fortress-like corporate structure to fend off would-be rivals who coveted the resources fiefdom he had built up.

Starting out in the 1970s with an Australian company called Otter Exploration NL, which merged in about 1978 with listed New Zealand firm Mineral Resources Ltd, Radford and Kennedy had by the early 1990s created a corporate labyrinth with complicated cross-shareholdings; holdings included significant controlling stakes in gold mines at Beaconsfield, Tasmania and in the Tanami desert in the Northern Territory, as well as the Martha gold mine in New Zealand's Coromandel, the Kupe offshore gas discovery in Taranaki, and the prospecting rights to the Pike coalfield.[3]

Radford was cautious and highly intelligent. In business he was autocratic and controlling; dissent was not welcomed and those who challenged his authority didn't last long. Among the people close enough to be invited to his pleasant home in Wollstonecraft, North Sydney, he was a witty and charming friend – at least until he fell out with them, as often happened. Over the years he had earned a reputation in the business community for falling short of accepted standards of good corporate governance,[4] and had accumulated some powerful enemies.[5]

The licence to prospect for coal at Pike had come into Radford's web in late 1979 when geologist Terry Bates, exploration manager for Otter Minerals Exploration – a subsidiary of the Tony Radford-dominated

Mineral Resources New Zealand – acquired it for the company. The presence of a thick seam of coal in the mountains inland from Punakaiki had been known about since 1911, but it wasn't until 1946 that the first geologist, Harold Wellman, set foot on the coalfield. With no road access into the formidable and isolated terrain, no one had ever attempted to mine it. Another three decades would elapse before the first commercial evaluation of the field was carried out, by Australian company Robertson Research (Australia) Pty Ltd.[6]

In collaboration with a bright young Canterbury University geology student, Jane Newman, and two fellow PhD students, including Newman's husband Nigel, Bates undertook extensive mapping and sampling of the coal seam in the early 1980s, often enduring atrocious weather and having to bash through thick scrub on the Paparoa tops until tracks and campsites were established.[7]

After four years of active study, including the drilling of six exploratory boreholes into the coal seam from the rugged surface, the Pike licence was put to one side. Bates got involved with the group's gold interests, and New Zealand was in the midst of political and regulatory upheaval that saw the state-owned land overlaying the Pike coalfield shifted from the Forest Service to the newly formed Department of Conservation. In 1987 the new Paparoa National Park was formed, adjacent to and partly overlaying the Pike coalfield.

The existence of the coal – and other valuable minerals – was taken into consideration when the park was being created, and only a small section of the Pike licence fell within its boundaries. However, having a landlord whose mandate was the preservation of New Zealand's indigenous flora and fauna and the protection of its places of wild scenic beauty didn't help increase the appeal of the Pike prospect to prospective mining partners.

Periodically one of Radford's executives would pick up the Pike licence and look at what could be done to extract value from it, then put it back on the shelf.[8] In 1988 a reshuffling of Radford's empire saw the licence shifted into the folds of New Zealand Oil & Gas (NZOG), one of a clutch of companies he and Kennedy had formed in the early 1980s and floated on the New Zealand and Australian sharemarkets. It was thought that NZOG, with its focus on hydrocarbons, might be able to find a partner capable of developing and extracting value from the Pike coalfield.

But no one could be tempted. NZOG geologist Roger O'Brien and senior executive Graham Mulligan went to Japan and Korea in the late 1980s to look for 'farm-in' partners; they were about to board a plane to China to scour for potential partners there too, but Tiananmen Square had become a sea of death and trauma, and they were prevented from travelling. They talked to companies in India and Indonesia.[9]

The only flicker of interest came from the Japanese giant Mitsui Mining. In 1992 one of Mitsui's most respected and experienced engineers, Masaoki Nishioka (known by all as Oki), was running a trial at the New Zealand government-owned Strongman underground coal mine on the West Coast to demonstrate the effectiveness of hydro mining – a method of extracting coal from the face with a powerful jet of water, which had been developed in underground mines in Japan and Canada.

One day O'Brien knocked on Nishioka's door at the Ashley Hotel, the Greymouth establishment favoured by professionals and businesspeople visiting the West Coast's major town. O'Brien's view was that the Pike coalfield ought to be developed first as an opencast mine to extract the shallower coal, with underground workings established later on when the money from early sales was flowing. He wanted to find out from Nishioka if hydraulic mining could work in an underground operation at Pike.[10]

It was a productive meeting: Nishioka proposed drilling seven boreholes to find out more about the geology of the coalfield. Back in Japan, he secured the funding needed to pay for the work, which was done in 1993.

Nishioka concluded the coal had valuable characteristics, but recommended to the Mitsui board that it shouldn't get involved in mining at Pike. With a soaring escarpment and a national park on one side and a densely forested wilderness on the surface, access would be difficult and expensive, he advised.

There were other factors that concerned him too. The coal extracted from the boreholes contained so much methane that the gas had bubbled out of the samples. One highly experienced man commented that it was the gassiest coal he'd seen in 40 or 45 years of mining.[11] Any underground mining operation at Pike would have to very carefully manage the methane, a gas which is explosive within the range of 5 percent and 15 percent of air and poses one of the greatest risks in underground coal mining.

Another deterrent was that the upper layer of the seam was high in sulphur, which would make it useless as a coking coal for the steel industry. Nishioka calculated that Pike would produce only about five or six million tonnes of saleable coking coal – not enough to justify the significant investment that would be required to develop the mine.[12]

New Zealand Oil & Gas's task of finding a partner to help it develop the Pike resource wasn't helped by the low coal prices prevailing at the time. The influence of China's economic revolution had not yet filtered through to international commodity prices; the price of coking coal through the 1990s was stuck stubbornly at around US$40 a tonne. Early calculations suggested it would cost most of that to get the coal out of the ground and transport it to international markets.

By 1997, the company was running out of momentum. Nothing much was happening with Pike – although by then the company had managed to get a full mining licence for the field and O'Brien had overseen some rudimentary pre-feasibility work. There was little oil and gas exploration work going on, and the Kupe gas discovery was taking years to bring into production. The Wellington head office was shut down and NZOG, including Pike River Coal Ltd, the subsidiary company formed in 1982 to hold the Pike licence, was shifted to Sydney, where it shared offices with another of Radford's companies, Pan Pacific Petroleum.

New Zealand Oil & Gas was a shadow of its former self: gone were the grey-haired executives and geologists who had formed the company's backbone. Its long-standing exploration manager Dave Bennett had been sacked by Radford in the mid '90s.[13] O'Brien was fed up with the lack of exploration activity and left. Graham Mulligan, the commercial general manager, had stepped down in 1994, although he remained a director. Terry Bates and Dave Kennedy, the geological brains behind Radford's wider minerals empire, had also split with Radford in the early 1990s. The company no longer had anyone in-house with geological or exploration knowledge of Pike.

Of the original Wellington-based NZOG executives only Gordon Ward, an accountant who had been hired by Mulligan in the late 1980s, moved to Sydney and remained on the staff.

Ward was from a dairy farming family. He had been an auditor for Coopers & Lybrand in Palmerston North – coincidentally, Radford's hometown – before joining the NZOG payroll. Like Radford, he was hard-working, disciplined and focused. He was a capable accountant, but to many of his colleagues his ambition exceeded his abilities. Robotic and rather colourless, he was certainly not leadership material. 'There was very little personality there. It was all business, all day,' recalls former NZOG and Pike River Coal director Peter Liddle.[14]

As the Pike prospect had been shopped around to potential partners without success for years, it had become obvious that, to get any value out of the asset, more investment would have to be put in. But NZOG was in no mind to invest hard cash into the project – its board of directors wanted to focus on oil and gas exploration, not get sidetracked on to a peripheral project.[15] Capital would have to be found from other investors to help pay for the work needed to progress Pike along the feasibility and regulatory pipeline.

By the late 1990s Ward had been elevated to the role of finance manager. He had no expertise in coal mining, but he picked up the challenge of securing finance and obtaining regulatory consents for the Pike project with enthusiasm. Those around him would come to call the project 'Gordon's baby'. It would catapult his career from the tedium of accounting to the entrepreneurial excitement of a brand new state-of-the-art coal mine.

The obscure coalfield on the South Island's West Coast, which few outside the small community of coal miners and geologists had heard of before, soon became the subject of headlines in the business press. A coal mine at Pike would be quickly profitable, Ward predicted in 1998.[16]

He went off to the coal markets of North Asia and came back reporting keen interest from potential buyers.[17] Enthusiastically at his side was consultant geologist Peter Gunn. Gunn had been interested in Pike since 1984 when, as district geologist for State Coal Mines, he had tried without success to convince his bosses to buy the licence. Brought in as an adviser by Roger O'Brien in the mid 1990s, he had produced a short marketing study that suggested a 'softly softly' approach to mine development that would enable further evaluation of the geotechnical aspects of the field and the volume of gas contained in the coal.[18]

In late 1998 NZOG began titillating the interest of sharemarket investors with talk of a public listing of the Pike project. The public discourse was infused with increasingly ambitious numbers: back in 1993, Mitsui had calculated that a total of five or six million tonnes of high quality coking coal could be mined from Pike, but by 1998 Ward was talking of 30 million tonnes.[19]

Consultant reports supporting the commercial and technical feasibility of developing Pike into a working coal mine started piling up. By 1998 any thought of using opencast mining methods was long gone: the notion of stripping such glorious untouched wilderness to dig out coal was certain to arouse insurmountable opposition, not only from the Department of Conservation but from a public that was increasingly interested in environmental protection. Gunn knew an opencast mine would enable more coal to be extracted 'but we considered that the environmental cost would have been too high'. Opencasting would have left a lasting impact 'and we would not have been happy with that'.[20]

In late 1997 Gunn – on behalf of NZOG – asked New Zealand mining consultant Dave Stewart to whip up a quick desktop study of an underground mine at the Pike coalfield capable of producing 500,000 to 600,000 tonnes a year. With scanty geological information to go on,

Stewart came up with a preliminary scheme within just a few weeks: he emphasised it was merely a 'first pass' report and that much more detailed investigation and exploratory drilling would be required.[21]

Stewart heard nothing more from NZOG. Two years later he found out that an Australian consultancy, Minarco,[22] had produced, on behalf of NZOG, a 'final feasibility study' that drew on his own elementary report.

The basic mine design drawn up by Stewart – a long stone tunnel to the coal seam and extraction using a high pressure hydro-mining machine – was replicated in the Minarco report. But there was a key difference: Minarco predicted a rate of production of more than one million tonnes of high quality coking coal a year, double the volume assumed in Stewart's report. Not only would there be exceptional productivity at the proposed mine, according to Minarco, but it would produce a healthy 29 percent rate of return to its investors.[23]

By then those investors included Ward's family, and senior people associated with Minarco itself. For the Pike project to make progress to the point where a wider circle of investors would be prepared to back it, Ward had needed more information, but assembling that information came at a price that NZOG, still the majority owner of Pike, didn't want to pay. The Minarco study was budgeted to cost just $100,000, but without the money to pay for it Ward suggested the consultancy take shares in Pike in exchange for a suite of studies and feasibility work.[24]

It was a sweat equity deal: investors associated with Minarco would get a 25 percent shareholding in Pike River Coal Ltd, and in return the consultancy would come up with a 'bankable' feasibility study – one with enough detail to satisfy lenders and investors – plus hand over $250,000 in cash. Minarco director Graeme Duncan, an Australian mining engineer, was one of those who took shares, along with fellow Minarco man David Meldrum. Duncan also became a director of Pike River Coal Ltd.

Duncan saw significant merit in the Pike project.[25] During his years as a consultant he had reviewed every New Zealand underground coal mine and regarded them as rather 'low-tech' in comparison with Australian mines. By introducing reliable, high-capacity equipment and 'a more Australian-style level of management, maintenance and systems', he was confident Pike could produce twice the amount of coal per year than had been assumed in Dave Stewart's report.[26]

Gunn was also a keen supporter of the project and remained heavily involved – analysing the geology, figuring out how coal would be transported from the remote mine to the marketplace, and collecting information on the pristine natural wilderness that overlay the Pike seam. NZOG asked him if he, too, would take shares in Pike in exchange for some of his consulting work, rather than being paid in cash.[27]

Ward, now on the way to becoming the chief executive of a new publicly listed coal mine, was proving to be just the man to get the Pike project on to its feet after two decades going nowhere. And while many of his former colleagues at New Zealand Oil & Gas had considered him bumptious and overconfident, Duncan and Gunn thought highly of him and admired his drive and commitment.

Precisely what role Tony Radford – who controlled NZOG as its chair and chief executive, as well as chairing the board of Pike River Coal Ltd – took during this time is unclear. Given his characteristic desire for tight control over all that occurred within the companies under his command, Radford would not have given Ward unbridled freedom to pursue the Pike project. 'Radford's style was to give people opportunity, scope and loads of rope, but he would always control from the back,' says Tony Frankham, who was involved with Radford from the 1970s, until they fell out in 2000.[28]

Radford perhaps saw Ward as a safe pair of hands, and a man he could easily relate to – after all, both were accountants, driven by numbers rather than relationships. And Ward would have known better than to challenge the authority of his boss: he had seen Radford cut adrift plenty of highly capable people over the years.

In any case, Ward was delivering the goods: he'd put his own family's money on the line, and had secured investors and a positive public profile for an asset that had languished within NZOG for years. He was also a determined advocate in the torturously slow process of gaining agreement from the Department of Conservation to allow the mine to be developed on the department's land, and extracting the resource consents required from the various local authorities.

The Pike licence was finally being infused with value, and it had begun to look as if Radford's empire might finally get a financial return on the asset after 20 years of ownership. And it was largely down to Ward's single-mindedness and zeal.

Ward could not have recruited a more energetic and dedicated advocate for the Pike project than Tony Kokshoorn. A wiry, voluble man whose words burst forth from his mouth in torrents, Kokshoorn wanted Pike to get going just as much as Ward did. Like most of his peers who had grown up in the Grey District, Kokshoorn had been raised in a mining family. His father Frank had emigrated to New Zealand from The Hague in 1952, moved to the West Coast timber town of Ruru and married a Coaster, and then ending up working underground at the Liverpool mine not far from Greymouth.

But as the young Kokshoorn grew up during the 1960s and '70s, the industry that had sustained the region went into steep decline. The Maui natural gas field off the coast of Taranaki was pumping cheap energy through the New Zealand economy, undermining the viability of coal. The colliers that used to queue up at the Port of Greymouth to load coal to fuel the New Zealand economy gradually stopped coming, and mines started closing.

Not only was the coal industry shrinking, but massive government restructuring during the 1980s wiped out jobs in the railways, telecommunications and postal services. The forestry industry suffered too, as mills that had relied on local rimu, beech and kahikatea lost their source of supply: the political tide had turned against the logging of the West Coast's indigenous rainforests.

The upshot was what Kokshoorn describes as a 30-year recession, during which the region's biggest export was its people. Between the late 1960s and 2000, the West Coast lost a quarter of its population.

Kokshoorn, who made his career in business as a used car salesman, had been an elected member of the Grey District Council for a year when Gordon Ward turned up at a council meeting in 1999 with the news that he was thinking of opening up a new coal mine in the Paparoas that would employ 150 people. Kokshoorn – known to his local constituents as 'Koko' – was instantly behind it. 'We hadn't seen anything that would employ 150 of our people for a long, long time.'[29]

Ward visited Greymouth regularly, pressing the local bureaucrats for the necessary resource consents, and negotiating with the Department of Conservation (DOC) for access rights over its land for mine roads and surface infrastructure. Strategic sponsorships of local events, such as the Moonlight Triathlon, helped cement goodwill towards both the Pike project and Ward, who participated in the annual race.

Kokshoorn was Ward's willing and able local support crew. 'The one thing on my mind was economic development for this area,' he recalls. When it became necessary to apply pressure to Wellington politicians whose officials placed hurdles in the way of the project, Kokshoorn was happy to take his lean frame and unvarnished advocacy to the capital and bang on tables.[30]

In 2004 – the year Kokshoorn was elected mayor of the district – the minister of conservation, Chris Carter, approved the long-sought access agreement allowing the mine to be developed. Pike and DOC then negotiated a set of detailed rules under which the mine could operate: the width of the road that would be built through the forest to the mine site; provision for emergency exits from the underground workings into the national park; protection against surface subsidence; the size of the ventilation shaft site on the side of the mountain; the amount of annual compensation to the landlord.

In time, some of the rules that Ward negotiated with DOC were found to be far too restrictive and had to be renegotiated. But Pike chair Tony Radford and director Graeme Duncan were clearly satisfied with the deal because they affixed their signatures to it as the company's representatives on October 24, 2004.[31]

Ward had done well. With a little help from Koko, he had got the proposed mine project over its biggest hurdle so far.

Despite the impressive progress Ward had made, by 2004 the project still consisted only of him and a covey of consultants, some of whom had shares in the project they were assessing and advising on.

By then Ward had been elevated to general manager of New Zealand Oil & Gas, and was based in a re-established Wellington office, but Pike remained little more than a sparkle in the company's eye. It had a mining licence, an access agreement and resource consents, and a mounting pile of reports from consultants, each building on those written previously and each adding to the increasingly cheerful view of the proposed mine's prospects. But it was still far from being a working coal mine, despite the confident early predictions of Ward and Minarco that it would be producing over a million tonnes of coal a year by the early 2000s.

Nevertheless, optimism reigned. International coal prices were on the way up, and it was claimed that, with the very best machinery and mining methods, Pike ought to be able to achieve stellar levels of production within two years of opening.

The proposal's weaknesses were redefined as strengths. Minarco advised in its 2000 feasibility report that Pike River Coal Ltd's 'absence of coal mining experience is expected to be an advantage as there is no preconception on management issues'. The company's unfamiliarity with the coal industry would not be a problem 'provided relationships with key advisers are appropriately structured and managed'.[32]

NZOG's first preference was to engage a specialist contractor to do the mining, prepare the coal for market, and manage the transportation from the mine to the customers.[33] After all, it had no expertise or project management teams that could be mobilised to the West Coast to begin the complex high-stakes job of developing and running an underground coal mine in difficult terrain, where hazards such as methane, outbursts – violent ejections of coal and gas from the seam – and cave-ins would have to be vigilantly managed.

But a contractor could not be found. Tenders were called and a couple of large Australian mining companies put in bids, but neither was satisfactory.[34]

New Zealand Oil & Gas, as the majority owner of Pike River Coal Ltd, now decided the only way forward was to begin from scratch and become a coal miner. It was to be a true start-up: everything from safety management systems through to employment contracts, mining machinery through to office equipment, worker training systems through to managers' job descriptions would have to be designed or purchased from a standing start.

Set in a wild and pristine natural environment, and run by people who would bring fresh ideas and high expectations to the hidebound world of underground coal mining, it would be the greenest of greenfield operations.

The man appointed to drive it forward was a rotund and charming Australian named Peter Whittall. Graeme Duncan was one of those who interviewed Whittall for the job of mine manager, and judged him to be a 'stand-up guy. His experience was excellent. His understanding and approach to safety was an important part of why he got the job.'[35]

Whittall arrived in Greymouth in February 2005 to confront a formidable undertaking. There were 'no roads, no nothing. It was just a mountainside and a feasibility study,'[36] he would later recall. A mining engineer with qualifications as a mine surveyor and mine manager, Whittall had spent 24 years with international mining giant BHP in Australia. He soon realised what he had lost in leaving a large multinational and joining a start-up. At BHP he'd had access to 'any system you want and any person you need'. He'd been a member of internal email forums where he could seek advice from up to 900 colleagues around the world. At Pike, 'I had literally no one. I had no systems. We didn't even have a payroll system. NZOG paid us but we had nothing. ...We had no HR systems, had no forms. I got my own IRD number when I started ... and we went from there.'[37]

From this blank sheet, he would make a coal mine.

TWO
Great Expectations

Greymouth sits at the mouth of the Grey River and spreads south in a grid of mostly flat, plain streets on either side of State Highway Six. The bush descends from the hills behind to form the town's soft, lush eastern boundary, and the Tasman Sea hammers at the long stony shoreline to the west. In winter a hard wind from the mountains – known simply as The Barber – often sweeps down the river, slices through the town, and leaves a thick white veil of mist across the Paparoas.

The town and its neighbouring settlements nurse a history of suffering. Mining tragedies in West Coast coalfields have taken 430 lives since the industry began there in 1864. The treacherous Grey River bar has periodically defeated the efforts of fishermen to reach safe harbour. Memories are fresh and raw of the 1995 Cave Creek tragedy, in which 14 people – 13 students from Greymouth's Tai Poutini Polytechnic and a Department of Conservation worker – died when a precipitously high viewing platform in Paparoa National Park collapsed. Until the advent of the Great Wall of Greymouth – the colloquially named flood wall built in 1990 – the Grey River repeatedly topped its banks and wreaked economic and emotional havoc through Greymouth's commercial centre.

However, Peter Whittall and his family arrived from their hometown of Wollongong, New South Wales, to a region on the up. Property prices were rising after a long period of stagnation. Local dairy farmers were coming into a period of prosperity, and the West Coast's dairy cooperative was working on a major upgrade of its milk powder factory at Hokitika, 30 minutes south of Greymouth. Solid Energy's Spring Creek underground mine was finally producing coal after years of difficult stop–start gestation. Multinational mining company OceanaGold was developing a large open-pit gold mine at Reefton, an hour up the Grey Valley. Attractive shops and tasteful cafes were opening up in downtown Greymouth.

And there was great excitement that the Pike River coal mine, with its promise of 150 or more well-paying jobs, looked as if it would at last go ahead. There was a sense that Pike, in particular, would underpin a new era of economic stability and well-being for the Coast.

Peter and Leanne Whittall threw themselves and their three children into their new community with gusto. They bought a rambling old villa on the corner of Heaphy and Shakespeare Streets, just a block back from the main thoroughfare through town. They repiled it, did up the kitchen and gave the house a fresh coat of paint. They joined the parish of St Patrick's, where Peter became a regular reader from the Old or New Testament, and the children went to the local Catholic schools. Despite his ample girth, Peter was an enthusiastic and competitive squash player. Before long he was appointed chair of the local industry group, Minerals West Coast Trust. Leanne taught at St Patrick's School.

Before he could get on with developing the new coal mine deep in the Paparoas, there were some basics to attend to. Along with environmental manager Ivan Liddell, who was hired a week after him, Whittall had to find office premises in Greymouth for the fledgling business, buy laptops and desks, get the phones connected, organise corporate apparel, and hire and build the team of skilled mining personnel that would be needed to develop the mine.

Within a few months, a skeletal staff had been established. Australian Denise Weir was recruited out of the Middle East oil industry, where she specialised in organisational development, to take up the role of human resources manager. Australian electrician Tony Goodwin became Whittall's right-hand man as engineering manager. Both arrived in the depths of the West Coast winter. New Zealand mining engineer Guy Boaz was hired as technical services manager.

Having been involved in the design and initial development of BHP's new Dendrobium underground coal mine in Illawarra, Whittall seemed suitably qualified for his new role. He approached the daunting challenge ahead of him with supreme confidence and optimism. He and Gordon Ward, who continued to serve as NZOG's general manager in Wellington, appeared to get on well and complemented one another's skills. As an accountant, Ward was completely reliant on Whittall's technical mining knowledge, and Whittall respected Ward's role as the money man. Where Ward presented as drab and haughty, Whittall was lively, clever and charming. He had the gift of the gab, with a particular talent for explaining the technicalities of the coal mining industry to potential investors.

When Whittall arrived in Greymouth in early 2005 there was still no guarantee the Pike project would proceed, but within five months he and Ward had co-authored a detailed document aimed at convincing the Pike board – still effectively a subset of the NZOG board, chaired by its veteran boss Tony Radford – to give it the green light and proceed towards a public sharemarket float to raise the necessary capital. Ward and Whittall based the document on the feasibility work completed by Minarco. Among their conclusions they stated: 'This operation will break new ground in New Zealand coal mining for productivity levels and annual production targets.'[1]

Pike would be the second largest export coal mine in New Zealand, carving out between one million and 1.4 million tonnes of coking coal every year – half as much as the entire export output from government-owned Solid Energy's various West Coast mines. This bold projection was a far cry indeed from the long-forgotten report by New Zealand mining engineer Dave Stewart in 1998, which assumed an annual output of just 500,000 to 600,000 tonnes a year.

Ward and Whittall were committed to a mine that had minimal impact on the environment; they were aiming for an 'exemplary' record, and proposed to manage the site in a way that would produce an overall improvement in the area's ecology. There would be comprehensive pest management and weed control programmes; the habitat of species such as kiwi and whio, the nationally endangered blue duck that depends on fast-flowing river environments for survival, would be enhanced.

A 'pragmatic and supportive' relationship had been established with the landlord, the Department of Conservation, Whittall and Ward wrote. Despite the long delay in winning an access agreement to mine under conservation land, DOC was certainly cooperative. Over time the department would agree to every request put forward by Pike to drill exploration boreholes from the forested surface, and every application for variations to the access agreement.[2]

A road would be built from the Grey Valley to the confluence of White Knight and Pike Streams, deep in the Paparoa Range. From there, a tunnel would be punched through the mountainside, driving uphill through rock to reach the Brunner coal seam after 2.3 kilometres. Roadways through the seam would be carved out by modern continuous mining machines with large rotating steel drums equipped with cutting teeth. The bulk of the coal would be extracted with the use of a hydro monitor, which would shear the coal from the face with a laser-like jet of water. It was thought that this method would be particularly suited to Pike's thick and steeply dipping seam. By this time 27 exploratory boreholes had been drilled into the Pike coalfield. They provided the basis of the company's understanding of the geological conditions the operation would encounter.

Coal extraction would proceed cautiously on account of DOC's concerns that mining could cause subsidence of the land above. A small trial panel would be mined in the north-western area of the mine before extraction began in earnest.

A ventilation shaft more than 100 metres deep would lead from the forested mountainside down into the mine's underground workings, providing a ventilation circuit for air contaminated by methane and other gases. In an emergency the workers would be expected to climb up the shaft if, for some reason, the main tunnel were blocked. Use of the shaft as a secondary egress was to be merely temporary, until more practical walk-out exits were built as development proceeded towards the trial mining area.

Demand for the low ash coking coal that Pike would produce was strong, Whittall and Ward claimed. Two major Japanese steelmakers had signed conditional long-term supply contracts, Brazilian and Korean steelmakers had shown interest, and Indian coke maker Saurashtra Fuels had signed up to buy Pike coal for the entire life of the mine. Indeed, Saurashtra was so keen on Pike coal it bought part of the company: in September 2005 it invested $17 million in return for a shareholding of 10.6 percent.

With Saurashtra on board as a major new investor, the Pike board made a formal decision to proceed. NZOG and other existing investors also committed $23 million.

A few months later the project won the backing of another big Indian coke maker, Gujarat NRE, which invested $20 million and also entered into a long-term contract to buy Pike coal. Between Saurashtra and Gujarat, Whittall and Ward had managed to secure buyers for more than half of the mine's projected annual production – an excellent result for a project that had yet to produce a single bucket of coal.

Whittall was a persuasive and positive front man. In November 2005, when work on the road to the mine had barely started and he had recruited a senior project management team of only four people, he told the Australasian Institute of Mining and Metallurgy's annual conference there was 'minimal remaining development risk' associated with the mine. The geology of the area was 'well understood' and the planned mining techniques proven. While there were challenges, 'Pike River coal mine will have the advantage of not only being a greenfield site but also a greenfield organisation.' It had 'a planning sheet devoid of past operational and management sub-optimal decisions (baggage!).' It would apply state-of-the-art technical and management systems, and had the opportunity to recruit and develop a workforce with 'aligned goals'.[3]

Outside experts chipped in with reports endorsing a rosy view of Pike's prospects. In mid 2006, consulting firm Behre Dolbear Australia addressed important questions that had been raised by bankers. What was so different about Pike that it would not face the delays and disruptions that had been experienced in mines elsewhere (a reference to Spring Creek mine, a project that had promised to be the mining industry's great new hope but had turned into a nightmare of cost overruns and production delays)? The lenders also wanted to know the basis upon which Pike would be able to achieve its ambitious forecast production tonnages. Were the predicted volumes achievable and realistic?

Yes, advised Behre Dolbear's managing director John McIntyre, a Sydney mining engineer. Among the reasons for confidence was that the Pike River management and engineering team had developed their skill and approach in the highly competitive and competent underground coal industry in Australia, he wrote. Pike would achieve its ambitions by using the best technology and software and 'doing what Australian mine operators do best – maintaining a high level of systems management to optimise the utilisation and availability of equipment'. Pike would have higher capacity hydro-mining equipment than nearby Spring Creek, which McIntyre advised had been held up by methane and ventilation problems, with large slugs of gas being emitted from mined-out areas and workers forced to stand down while the mine was de-gassed.

Spring Creek was also burdened with complex geology, he noted. Unexpected underground features had caused significant downtime and loss of production. Pike's coal seam, by contrast, was 'relatively simple'. And Pike would have superior, purpose-built mining machinery.

McIntyre judged that Pike's annual output would exceed Spring Creek's by more than 50 percent.[4]

By 2007, the long-signalled initial public offering (IPO) of shares in the proposed Pike River mine was ready to go ahead. Ward and NZOG had been promising it to the investment community since the late 1990s, but there was now tangible evidence that there really would be a mine to invest in. The access road had been constructed through farmland and dense forest to the tunnel site, curving beautifully and sensitively around ancient rimu and beech trees. The first shots had been fired in the stone face near the junction of White Knight Stream and Pike Stream back in September 2006, marking the start of construction of the 2.3-kilometre tunnel to the coal seam; by May 2007, men and machines had progressed almost 800 metres under the Paparoas. Brand new mining equipment, including a roadheader and two continuous miners, had been ordered from Australian company Waratah Engineering. Orders were in for loaders and specialised mining vehicles designed to shift men and materials in and out of the mine.

Greymouth was starting to reap enormous spin-offs from the project, and the town had in turn taken Pike and its new recruits to its heart. Human resources manager Denise Weir worked hard to establish links with local schools and Tai Poutini Polytechnic to ensure the new mine would have a flow of young people interested in a career in the underground coal industry. She built relationships with local politicians and employers who could help the company navigate any regulatory and political obstacles to the recruitment of the skilled foreigners needed for specialist roles. And she worked assiduously to ensure that those who were recruited from Australia, South Africa, the United Kingdom and beyond were well supported when they got to the district. Weir – who came to regard the West Coast as her spiritual home – was left in no doubt about the importance of the Pike project to the region, or how intensely interested the locals were in its progress: when she walked to work at the small Greymouth office that Pike used as its headquarters in the early months, people would come out of their houses to greet 'the Australian lady' who had come to work for the mine.[5]

As the project ramped up, there would be work for dozens of West Coast contracting businesses – not just the big operators such as Ferguson Brothers, who won the job to build the new road to the mine site, but for sundry tasks: construction of the miners' bathhouse and mine administration buildings, earthmoving, signwriting, plumbing, engineering, laundry, road transport, helicopter transport, and more. The new project gave local contractors the confidence to invest in new equipment. And the community benefited from Pike's beneficence as the company bestowed sponsorship money on dozens of organisations, ranging from the Ikamatua Golf Club to the Mawhera Young Writers' Club and the Greymouth Trotting Club.

According to the May 2007 prospectus, large volumes of coal would be pouring out of the mine and into the booming international coking trade by March 2008, just ten months away. By 2009 it would be producing a million tonnes annually.[6]

Conditions for the IPO were highly favourable. The price of coking coal, in the doldrums for decades, had more than doubled in 2005. And although prices had since eased back, they were forecast to remain considerably higher than their historical average. Almost three decades of double-digit economic growth in China was creating massive demand for steel to build skyscrapers, sprawling megacities and modern transport systems. Demand from the Indian economy was also booming. Coking coal was a hot commodity, and hunger for it was predicted to increase. As Ward had enthused to financial journalist Jenny Ruth in late 2004, it was a 'fantastic time for coal producers and soon-to-be coal producers'.[7]

After the IPO and sharemarket listing, Pike would be the only publicly listed coal mining company in New Zealand. Its entrée into the NZX would give small investors the chance to participate in the global resources boom. The prospectus boasted that the company had 'quality leadership and governance', although it was acknowledged that neither it nor the independent experts advising it had any operating history on which to base their projections of future costs, capital requirement, and revenue and production targets. In a sweeping caveat, the prospectus noted that its projections were 'necessarily estimates' and 'sensitive to a number of assumptions and risk factors'.[8]

Among the listed risks were those associated with all underground coal mines – gas, spontaneous combustion in the coal seam, roof collapse, fires, windblast and geological surprises. Despite the disastrous methane- and coal-dust-fuelled explosions that had killed so many miners in the region in previous decades, investors were assured that 'gas in the Brunner seam is generally at low to moderate levels, giving Pike River confidence it can safely and efficiently mine this seam.'[9]

There would be ample ventilation, the prospectus promised, and in gassier areas it might even be possible to harness the methane to generate electricity, lowering the mine's running costs.

Behre Dolbear Australia provided prospective investors with a reassuring technical review of the Pike project, which it said was based on 'adequate' geological and geotechnical data, with 'appropriate and reasonable' projections of production tonnages and costs. Due consideration had been given to the management of risks such as gas, ventilation, spontaneous combustion and environmental damage. Behre Dolbear, too, covered its report in protective caveats, pointing out that Pike proposed to mine in areas where the 'geological, mineralogical and geotechnical characteristics are only broadly understood at this stage'. The accuracy of the conclusions Behre Dolbear had arrived at in its review of the project 'largely relies on the accuracy of the supplied data'. That data was supplied by Pike and its legion of consultants.[10]

Pike sought $65 million from new investors. Demand was overwhelming. The company scooped up $85 million from thousands of investors. Along with bank borrowings, it was thought to be more than enough to finish the mine development.

The IPO was also crafted to create strong incentives for the company's leaders. Ward – whose family by then held over 700,000 shares – was given 100,000 shares for free and allotted another one million partly paid shares for a down payment of one cent each. Whittall was allotted 800,000 partly paid shares on the same terms as Ward.[11]

To celebrate the company's arrival on the bourse, Ward presented the NZX with a mounted lump of coal,[12] even though the project had yet to strike the coal seam. By the middle of 2008, the value of Pike shares had increased from one dollar to almost $2.50, and Pike joined the élite ranks of the NZX50, the index of New Zealand's largest listed companies, closely watched by financial media, brokers and market analysts.

Pike began to gain a reputation as a showcase development, blessed by visiting ministers of the Crown and celebrated for its superior environmental management and modern methods. From a mere pinprick in the beautiful West Coast wilderness, it promised to produce over four billion dollars in wealth for the New Zealand economy over the coming years, and prove that mining and a high standard of environmental protection were not mutually exclusive.

Gordon Ward, the former auditor from Palmerston North, had left behind the daily grind of bookkeeping and become the chief executive and managing director of a Top 50 company. And Peter Whittall, the articulate Australian mining man, was right beside him, driving the project forward.

Over the celebratory din it was difficult to hear the quiet thrum of doubt.

THREE
Early Warnings

One side of Jane Newman's Christchurch office is stacked high with box files and ring binders crammed with scientific papers, correspondence, data and historical records relating to the Pike River coal resource. With her soft blonde curls and gold-rimmed glasses, Newman doesn't conform to the sterotype of a rugged geologist, the sort of person likely to be an expert on the geology of the Pike coalfield. But by the late 1990s, when New Zealand Oil & Gas decided to develop the asset that had been sitting uselessly on the books for so long, she had developed an unrivalled knowledge of the area.

Starting in 1980 as a PhD student, she had worked closely with Terry Bates on detailed mapping and sampling of the field. With Bates and other students of the resource, she had camped in the shadow of Mt Hāwera, which stands sentry over the Pike field, traipsed across steep, bush-clad terrain drenched by more than six metres of annual rainfall, and chipped out countless samples of coal with the aid of a rock hammer. It was work that called for a resilient temperament and a certain daring: one group of Bates' 'fieldies' had its camp flattened in wild weather and had to walk out via Pike Stream, an eight-hour trek, and an outgoing, fit young geologist named Richard Cotton had climbed down the precarious western escarpment to gather samples from the outcrop, where both the

Brunner coal seam and the deeper Paparoa seam run in visible bands of black for more than six kilometres.

Newman was alongside Bates and his team when the first six exploratory boreholes were drilled in 1983 from the top of the Paparoas down into the coal seam. The core samples drawn up were akin to a geological biopsy, helping to build an early understanding of the complex and layered ground. At the top lies an overburden of hard island sandstone, formed from sands deposited when the area was inundated by sea during the Eocene era, when the first whales were evolving. Then comes the rider seam, a thin layer of high-ash coal, and below that an 'interburden' of mudstone and sandstone. Next is the thick Brunner seam, the remnants of a peat mire that has been cooked and compressed over 45 million years. Within the Brunner seam there are also isolated pockets of sandstone, laid down by streams that ran through the ancient mire. Below the Brunner seam are the much older Paparoa coal seams, formed 70 million years ago, at the end of the age of dinosaurs.

Later in the 1980s, as leader of the Canterbury Coal Research Group at the University of Canterbury, Newman supervised Masters and PhD students investigating the geological structure of the Pike field and the industrial properties of the coal. The government was keenly interested in learning more about New Zealand's coal resources, and funding was available for scientific work if there were evidence of support from the coal mining industry.

By 1988, though, Terry Bates was no longer actively involved in Pike exploration, and the reshuffle of Tony Radford's minerals empire had seen the prospecting licence shunted into Radford's hydrocarbon exploration company, New Zealand Oil & Gas. For Newman this marked the end of a satisfying marriage of interests between the academic and commercial worlds. She quickly discovered that dealing with NZOG was a one-sided and often fraught business, in which she gave far more – in the form of research data and archival material – than she received – in the form of cooperation and financial support for her research programme.

The drilling of a seventh exploratory borehole in 1990 was typical of the new order. Newman gained central government research funding to drill the hole, having received a nod of support from NZOG, which indicated it would drill another hole of its own. Newman duly completed her drilling work, but NZOG then declared it wasn't willing to spend any more money at Pike, blaming uncertainties created by planned new legislation, including the Resource Management Act and proposed changes to New Zealand tax laws.

In 1993, when the mining giant Mitsui came to New Zealand to run a Japanese-government-funded drilling programme at Pike, NZOG recruited Newman's husband Nigel, who also worked with the Canterbury Coal Research Group, to guide the Mitsui men on a field trip of the site. Newman's group also hosted the Japanese representatives, including top mining engineer Masoaki Nishioka, at their laboratory, where they discussed Pike's geology in greater detail.

In a letter following the Mitsui visit, Newman made abundantly clear her reservations about NZOG's management of the Pike licence and its reluctance to invest in adequate research and exploration. 'May I suggest that, if you do decide to joint-venture with New Zealand Oil & Gas, you request that Canterbury Coal Research Group be appointed local geological adviser. We have been very concerned that the area has not been professionally evaluated except by ourselves and we would like to ensure that no mistakes are made by New Zealand partners during exploration related to such a venture. In the early years of the present company's involvement we felt that funds were spent on inappropriate activities (seismic) which did not increase understanding of the resource.'[1]

Newman continued to oversee research into the Pike coalfield until the Canterbury Coal Research Group was disbanded later in 1993, and NZOG continued to benefit from the group's large bank of knowledge. Indeed, virtually all the research undertaken at Pike between 1990 and 1993 was done by Newman's group. NZOG's promises to help share the costs came to little.[2]

For the remainder of the 1990s Newman's involvement with the Pike River coalfield was mostly limited to responding to requests from NZOG or its representatives – particularly its consultant geologist Peter Gunn – for data and documents from her enormous archive of information on the coalfield. She gave it freely.

By 2001 Gordon Ward was in the thick of negotiations with the Department of Conservation over the access agreement needed for the mine to proceed. Among the department's concerns was the potential for acid mine drainage, one of the most serious environmental effects of mining and known to be a high risk in other parts of the Brunner coal seam. Acid mine drainage occurs when rock bearing the mineral pyrite is exposed to rain, groundwater and air as a result of mining. It causes run-off into creeks and streams that can be lethal to aquatic life. Experience from Solid Energy's Stockton opencast mine, north-east of Westport, where adjacent creeks had been rendered lifeless by mine run-off, showed the department was right to take the risk seriously.

In August 2001 Newman was asked by Peter Gunn to come up with a 'scientific and impartial study' into the potential for acid mine drainage, and he wanted it as soon as possible.[3] The objective was to head off any onerous conditions that the Department of Conservation might seek to impose on the access agreement. The key document with which she was provided to undertake this work was a copy of the final feasibility study prepared in 2000 by Minarco in return for shares in Pike River Coal Ltd.

Newman was immediately struck by how sketchy the feasibility study was, and how inadequate the geological knowledge underpinning it. For instance, it appeared to entirely overlook the existence of the rider seam, and the cross-sectional diagrams depicting the geological make-up of the coalfield were, to her mind, merely 'cartoons'. Over the years, her students had developed far superior insights into the geology of the coalfield than was apparent in this commercial study purporting to be the intellectual and financial basis for a capital-intensive new mine.

From her years studying the coalfields of the West Coast, Newman knew that a detailed understanding of geology was fundamental to the prospects of a successful mine. She had no doubt the Pike coalfield contained a unique resource that, if intelligently and safely exploited, had economic value. Indeed, she thought it had the potential to find its way, with proper research and marketing, into high value, niche coking mixtures suitable for specialised steels and perhaps even carbon fibre. But there was no evidence in the feasibility study that those promoting the mine had an understanding of Pike's complex stratigraphy – the interrelationship between the various rock and coal layers – and the science of analysing and modelling variations in those layers. Without an insight into the stratigraphy, as well as the geological complexity created by faulting, the developers of the mine would, she believed, fail to understand the nature of the risks they were taking on.

Newman made her views plain to Ward, writing to him in October 2001 about her concern at the lack of detailed geological information in the feasibility study. 'The one cross section which I have seen in the draft feasibility study is really only a schematic representation of the gross stratigraphy and structure, and completely inadequate as a basis for mine planning. It provides a misleading impression that the seam is uniformly thick and unaffected by sedimentary partings.'[4]

Later she wrote again to Ward and Gunn, pointing out that 'a successful operation is doubtful without at least short-term input from an experienced mine geologist (preferably someone with West Coast experience).' She made it plain she considered neither Gunn nor Graeme Duncan, whose company Minarco had authored the feasibility study and who was a key consultant to the project, as well as a director of Pike River Coal Ltd, had the necessary skills or knowledge.

Duncan retorted that Newman had been hired to help with the acid mine drainage (AMD) issue, and had overstepped her brief by criticising the geological content of the feasibility study. Undaunted, Newman pressed on. 'Please read the following carefully,' she rebuked Duncan. 'You are correct that my brief is officially limited to AMD/compositional studies, and that is in fact my preference. However, it is clear that some of the other parties unofficially regard my involvement as an endorsement of the operation as a whole, not just the AMD aspect, and I am uneasy about this in the absence of more evidence that the geological model is adequately supported.'

The lack of detailed information in the feasibility study, and the 'primitive character of the cross sections, does not inspire confidence,' she continued. 'I have a responsibility to make my concerns known to Pike River Coal Ltd in advance of any questions which may be put to me by DOC or other parties. … I would like to refer them to a mine geologist and neither you nor Peter [Gunn] fall into this category.'[5]

In response, Duncan freighted several boxes of further material relating to the feasibility study to New Zealand. It did little to reassure her: there was still no sign of the detailed cross sections that would illustrate that Duncan and the other advisers on the Pike project understood the geology of the area they were planning to mine.

Even the Department of Conservation staff, who were required to advise their minister on Pike's application for an access agreement, thought the project was light on detail. Hugh Logan, the director-general, regarded it as a 'back of the envelope' plan, with NZOG taking a 'just in time' approach to the provision of supporting information. The company's approach was in marked contrast to that of OceanaGold, whose proposal for a gold mine at Reefton was before the department's West Coast conservancy at the same time. OceanaGold would 'sit around the table, agree on a work programme, and deliver what was required on time'.[6] The two companies were like chalk and cheese.

Logan knew the department needed specialist help to assess the Pike application and supporting technical documents. Transiting through Wellington Airport in early 2000, he bumped into Murry Cave, a geologist he knew to have expertise in the coal sector. Logan put the Hokitika office in touch with him, and he was subsequently hired to provide technical expertise. It was Cave who, during a visit to the coalfield in early 2001, would identify acid mine drainage at the old mine adit – a horizontal tunnel to the coal seam, developed by Terry Bates in the early 1980s – and flag it as a significant risk the department needed to take into account.

Cave came to the task with detailed knowledge of New Zealand's coalfields: in the 1980s and early 1990s he had managed the division of the then Ministry of Energy responsible for information on New Zealand's geological resources, and had worked on the New Zealand Coal Resources Survey, a major study of the country's coalfields. He was co-author of the government-funded *Coal Resources of New Zealand* publication, and led an initiative to publish the scientific findings of the Coal Resources Survey as the *Coal Geology Report* series. He had also worked on detailed studies of West Coast coalfields and encouraged and financially supported university research, including work on the Pike River coalfield.

Like Newman, Cave was deeply unimpressed with the level of information that had been supplied in support of the Pike access arrangement. He expected it to consist of a large volume of technical material – including the results from borehole drilling, comprehensive geotechnical data, field mapping, analysis of the geological structure, and so on. Given the scale of the proposed development, he thought DOC would probably need to set aside a secure room at head office where he could work through the data. Instead, the material he was asked to assess was handed to him in an A4 envelope.

When he told Minarco's Graeme Duncan the documents did not provide an adequate basis for him to sign off, he was sent the 2000 final feasibility study and some borehole data. This was far from reassuring. The study didn't contain anything like the level of detailed analysis that Cave expected for a complex underground mining proposal, and the borehole data was sparse. He told DOC that Pike's promoters didn't seem to have adequate information on which the department could base a decision regarding access.

One of his concerns was that very little information had been provided on the methane content of the coal seam at Pike. Explosive gas – or firedamp, as the old miners called it – is a product of the same geological process that creates coal itself: decaying swamps, overrun by the sea or rivers, are buried, compressed and heated over millennia. Predominantly methane, the gas is held within the microscopic pores of the coal, which serve as a highly efficient storage system. When mining or drilling exposes the coal to the atmosphere, the gas migrates out. As methane is lighter than air, it rises. One of the key hazards in underground mining is 'layering' – the tendency for gas to collect along the roof of the mine, forming a path for flames to propagate along.

If methane collects in the air at concentrations of between five and 15 percent, it will explode in the presence of even a tiny source of ignition – the friction of metal picks on stone, for instance, or the spark from a machine. Even the small amount of energy produced by a cell phone, wristwatch battery or crushed aluminium can (among items banned in underground coal mines) can be enough to ignite an explosive build-up of methane.

Another risk associated with gas is outburst – the violent ejection of coal, gas and sometimes rock from the coalface. The West Coast had had painful recent experience of the dangers of outbursts, with three men killed in two separate incidents at the Mt Davy mine in 1998. The mine was subsequently shut down after it was decided the outburst risk could not be managed.

Underground miners assess the gassiness of their coal in terms of cubic metres per tonne: fresh samples drilled from the seam are sealed in special containers and the volume of gas released over time is measured. Although there are no strict definitions, miners tend to regard a coal seam with methane levels of more than six cubic metres per tonne of coal as highly gassy, from three to six cubic metres as moderately gassy, and anything under three cubic metres per tonne of coal as having low gas.

International standards established by the United Nations recommend that if the volume of gas is more than six cubic metres per tonne of coal, it should be pre-drained through drill holes well before workers are sent underground to mine it, a process that can take months or even years. From the minimal information provided about the proposed mine, Cave could see that the gassiness of the Pike seam ranged from just over three cubic metres per tonne of coal to well over ten cubic metres per tonne. Yet the Minarco feasibility study referred to an average of six cubic metres. Also of concern to Cave was that the gassiest coal was in the area near the Hāwera Fault, which would be under great geological pressure and was likely to present a risk of outburst.

Another worry was the way in which the emergency egress points in the proposed mine were designed and placed. These were to come out on the steep escarpment at the western edge of the mine, and would have had miners escaping into a precipitous environment where rescue would have been difficult. Cave thought it was important that an additional emergency egress was built before the mine reached the proposed trial mining area near the escarpment.

In late 2000, Cave advised DOC that there were critical gaps in the Pike database – the company hadn't drilled enough exploration boreholes, didn't have enough accurate gas data, lacked geotechnical and structural information, and didn't know enough about potential surface subsidence caused by mining.

Cave said more boreholes needed to be drilled from the surface to gather further information. But Pike's promoters told the department the plan was to gather the necessary geological information by trial mining, rather than by costly drilling.

The company's attitude was a red flag for Cave. 'The mere fact that Pike River Coal Ltd lacks the money to do all the work before getting access (and thus bankability) should send quite strong warnings about the project's economic viability,' he wrote to DOC. 'If they cannot get an investor confident to take the risk at this stage, then what is the risk of economic failure after they have got access? The last thing you want is a new road and mine adit which is then closed after a year or so because the mine is not viable!'[7]

Cave perceived that DOC was under pressure to make a favourable decision for Pike. When the company arranged what were purported to be technical workshops on the risks of subsidence and acid mine drainage, lawyers, local politicians and representatives of the deputy prime minister and minister of economic development were all in attendance.

Despite demands from Pike advisers Peter Gunn and Graeme Duncan that he provide them with a copy of his draft advice before delivering it to DOC, Cave didn't resile from his cautionary message. The feasibility study completed by Minarco was merely a 'useful basis' for further consideration of the mining proposal, he advised DOC in mid 2001. The overall quality and quantity of geological information it contained was 'limited'.[8]

The following year he appeared as an expert witness at hearings to determine whether Pike should be granted the resource consents it needed from the West Coast Regional Council and the Grey and Buller District Councils. He noted that the mine proposal had been developed on the basis of sparse exploratory drill holes that were, on average, over 500 metres apart, and in some cases over a kilometre apart. By comparison, the Spring Creek mine had drilled much more intensively – at spacings of between 150 and 300 metres. And despite this more diligent standard of exploration, that mine had been recently mothballed because of costly geological surprises.[9]

The experience at Spring Creek highlighted the high level of risk involved in trying to mine an insufficiently detailed resource, Cave warned. And while his client, the Department of Conservation, had no remit to assess the safety and technical viability of Pike's mine plan, the department could end up with a huge liability on its hands if inadequate and lightweight planning led to the operation failing or causing serious environmental damage.

As it turned out, Pike was granted the access arrangement it needed from DOC, as well as the resource consents from the local authorities.

After 2004 Cave had little further involvement but he continued to observe the progress of the Pike project with interest. When the company published the prospectus for its 2007 IPO, he read the document with a good deal more insight than the average investor. In particular, he was surprised at the casual reference in the document to the methane risk at Pike. After 48 pages of glowing testimony about expected coal production levels and the returns that would be achieved from lucrative international markets, the document stated that the Brunner seam had 'low to moderate' gas content, although high enough in some isolated areas to warrant pre-drainage.[10]

Yet Peter Whittall, the project's general manager, had spoken at a mining conference in 2006 about Pike's 'medium to high' methane levels, which would be 'difficult to control by ventilation means alone'. The intention was to drill holes in the seam to drain the methane down to a level of about three cubic metres per tonne, if possible, before it was mined.[11]

Investors in the IPO were being given a rather different picture. Cave thought the promoters were encouraging a false sense of security about the gas risk, and when he was approached for advice by prospective investors he made his concerns about this, and other aspects of the project, known.

When the media got wind of Cave's private comments, he was prompted to document his views for wider consumption. He wrote that the underground area Pike planned to develop to accommodate critical infrastructure for the projected 18-year life of the mine – a place known as 'pit bottom in coal' – was highly gassy, with an associated risk of explosive and potentially fatal outbursts. Pike was also projecting 'over-ambitious mining targets' and promising production levels that no company in New Zealand had ever managed to achieve from an underground coal mine. He pointed out the mine had to pass through the active Hāwera Fault, and said the geological make-up of the area was so complex that Pike couldn't be certain how much coal was actually there. He also noted that construction of the tunnel through rock to reach the coal seam was already proving to be slower than expected.[12]

The response from Pike's general manager was swift and angry. Whittall accused Cave of 'inaccurate, inappropriate and ill-informed' comment.[13] The independent experts who were advising the company were far better acquainted with the risks and benefits of the project than

Cave, he stormed. John Dow, Pike's board chair, also responded publicly, rejecting Cave's concerns and arguing that because Pike was a relatively shallow mine it wouldn't be troubled by outbursts. He added that the budget and expected timeframe for building the tunnel were in line with the company's knowledge of the rock conditions.

Harry Bell, too, harboured doubts about the Pike River mine proposal. Over a six-decade career in the underground coal industry, there was little Bell had not experienced first-hand. Along with his sister Nan Dixon, he had grown up in a mining family in the town of Rūnanga. At 15 he had taken his first job as a lowly ropeboy at the Liverpool underground mine, where he hauled miners' full boxes on to the weighbridge and unhooked the empties. Eventually he had become one of the top producing miners and climbed steadily through the industry's strict hierarchical ranks. He became a deputy at the Strongman mine, with legal responsibilities for underground safety, and then rose to the status of underviewer at Liverpool.

When an explosion ripped through the Strongman mine in 1967, Bell had led the second team of Mines Rescue Service volunteers to go underground. Beginning two and a half hours after the explosion, the men had searched and recovered bodies for 12 hours, before emerging exhausted at one in the morning. Four of the 19 dead were still underground, but it was too dangerous to go back in as there was evidence of fire. The part of the mine in which the men were located was sealed off until the fire was suffocated; three and a half weeks later Bell and others went back underground.[14] Two of the bodies were recovered, and two could not be found. Still today, they remain entombed in the mine.

After more than 20 years working underground, Bell attained the estimable role of mine manager, and after 30 years he was appointed regional inspector of coal mines. For two years in the early 1990s, before the mining inspectorate was unravelled in a wave of deregulation, he occupied the pinnacle of the industry as chief inspector of coal mines, a position of great mana and authority.

Bell was known throughout the industry as a tough and pragmatic straight shooter. There were few who were not aware of his standing as the man who had saved dozens of miners' lives at the Huntly West underground mine in 1992: as chief inspector he had ordered the place shut after workers noticed smoke emerging from an area of old workings. He was accused of overreacting to the risk but refused to buckle. Three days later the town of Huntly reverberated when an enormous explosion ripped through the mine. Thanks to Bell's dictate, the mine was empty and no one was killed.

In mid 2006 Bell was asked by a member of Pike River Coal's small and newly recruited management team to oversee the mining of several tonnes of coal samples from near the headwaters of the Pike Stream, where the top of the thick seam was visible at the surface and where the small adit had been built 26 years earlier for Terry Bates, under Bell's supervision.[15]

The instructions from Pike were explicit: Bell was to avoid taking samples from the top five metres of the coal seam, which were high in sulphur and therefore unsuitable for steelmaking. Bell and three workers from the contracting firm McConnell Dowell flew up to the site by helicopter each day for ten days. They used explosives to release the coal from the seam, and wheelbarrowed the sample coal out of the tunnel. They then filled dozens of 44-gallon drums with low-sulphur coal; these were carted away by chopper – slung three or four at a time – and then shipped to customers for their analysis.

Bell was not happy, and told Pike geologist Jonny McNee so. 'I said this is not a representative sample. A representative sample was from the floor to the roof of the seam. They were cheating.'[16] He worried that customers would enter into contracts on the basis of the low-sulphur samples, when he knew it would be very difficult for Pike to separate the high-sulphur coal from the low. It would be easy enough in an opencast mine to segregate the different coal types, but it would be an entirely different proposition in an underground mine. 'In an underground mine the extraction of top-class coking coal from high-sulphur coal is extremely difficult because if the mid-seam coal is taken out during extraction, the top coal will fall and the top sulphur coal will then contaminate the whole product.'[17]

The options being discussed by Pike for dealing with the problem seemed to Bell unworkable. One was to take out the good coal, and leave the high-sulphur top coal to fall to the floor and let it lie there. Bell advised that this remnant coal was likely to catch fire.

Another option was to take the high-sulphur coal out via a long pipeline, then clean the pipeline to remove all traces of sulphur and pipe the good coal out. He thought this would be very difficult, time-consuming and costly, and doubted it would be commercially viable.[18]

When he was asked by friends for his advice as to whether they should invest in the company's IPO, he advised them not to. He just couldn't see how Pike River Coal would ever make money.

While Newman, Cave and Bell harboured worries that the Pike project was being developed on the basis of inadequate exploration and flawed assumptions, serious fractures were appearing at the top of Pike River Coal Ltd. As the company worked towards getting its share float off the ground, Tony Radford had bowed to pressure from the organising broker, First New Zealand Capital, to relinquish the chair and bring in directors who were independent of Pike's major shareholder, New Zealand Oil & Gas.

Retired investment banker Denis Wood was invited to chair the board. He seemed a good fit: he was originally from the West Coast, his father had been a coal miner, and he had held senior roles in the financial markets – including acting as a member of the Stock Exchange surveillance panel that had probed Radford's own conduct in the late 1990s when, as chair of Otter Gold, he had disenfranchised the major shareholder, Guinness Peat Group.

Wood was well aware of Radford's reputation as a difficult character, and of the suspicion with which he was held by some in the financial markets. Nevertheless, he took up the role in April 2006. Also appointed was Wellington investment banker James Ogden. In common with the rest of the board, with the possible exception of Graeme Duncan, neither had any technical knowledge of coal mining, but they understood the requirements of the capital markets and the importance of sound and prudent governance. First New Zealand Capital was comforted by their presence.

By the time they took up their positions, it had already been decided that Gordon Ward would be appointed chief executive of Pike River Coal Ltd when it was floated as a publicly listed company. Doubts about the former auditor's competence for the role, and about his continued obligations as NZOG's general manager, soon began to niggle with the new directors: getting a greenfields mining operation up and running and floating it as a public company were major undertakings and there was no room for divided loyalties.

There was also rising concern about NZOG's level of financial support for its coal mining offshoot. The cash injected by the two Indian shareholders, Saurashtra and Gujarat, was being rapidly burned through as development of the mine project got underway. The project was being run on a shoestring, yet there was pressure from Pike management to sign off on major contracts for the supply of millions of dollars worth of new mining equipment.

Then, in the midst of Pike's efforts to raise capital from institutional investors and banks in advance of the IPO, NZOG itself launched a major capital-raising effort that cut across the efforts of its subsidiary. Trust between the two independent directors and Radford collapsed.

On December 8, 2006, Wood and Ogden resigned – Wood after just eight months on the board, and Ogden after a mere six months. Graeme Duncan, the long-standing director who had joined the board in 2000 as part of the shares-for-services deal with his then company, Minarco, resigned the same day.

None of the three publicly disclosed the real reasons for their resignations; the sharemarket was given the anodyne explanation that they had stepped down for 'personal reasons'. Duncan has subsequently stated he left because NZOG (through its directors on the Pike board, Radford,

Ward and Ray Meyer) had taken direct control of the capital-raising and IPO process and this had left him feeling unable to fulfil his obligations as a director.[19]

However, the truth of the breakdown was spelled out in detail in Wood's confidential letter of resignation. 'Put simply,' Wood wrote, 'my decision relates to governance issues, the relationship with NZOG and the potential financial exposure for the directors of Pike.' Wood attacked NZOG for usurping the role of Pike's board and putting the directors at personal financial risk by inadequately funding the company. 'The cash tank is now virtually empty yet the directors were being asked to approve project indemnities and other commitments that inevitably places them at personal risk.'

Wood referred to an interim funding proposal put forward by NZOG, describing it as unsatisfactory. Not only was the amount of money 'insufficient for the prudential funding of the project', but the fees demanded by NZOG were 'outrageously high'.

He was leaving the board with regret and a sense of disappointment given his high hopes for the Pike project. 'It is only due to my association with the West Coast and my desire to see this project to completion that I have stayed on longer than any prudent independent chairman would have done.'[20]

As soon as the three directors walked out, First New Zealand Capital quietly withdrew from the role of lead manager of the IPO. The mass resignation was an unmistakeable sign of serious dysfunction at the top. Nevertheless, two replacement directors were promptly recruited to fill the seats of the departed independents. John Dow, a retired gold mining executive who had spent much of his career with the multinational mining company Newmont Corporation, joined as a director, and took over the chair in May 2007. For Dow, who had retired at the young age of 59 and returned home to New Zealand to live, it was an ideal opportunity, with the added advantage that the mine site was a relatively easy drive from his home in Nelson.[21] Stuart Nattrass, who had worked in the international foreign exchange market and was looking to build a career as a professional director, took Ogden's vacant seat.

Neither man had any experience in the complex business of underground coal mining, and neither spoke to Wood or Ogden about the reasons for the previous directors' sudden resignations. Dow talked to remaining members of the board, Tony Radford and Ray Meyer, and concluded the reasons for the three-man walkout were 'relatively trivial' – the sort of thing that 'grown men resolve in an amicable fashion'.

For his part, Nattrass was comforted by the knowledge that Dow, an experienced mining man, would chair the board, and by the intelligence and integrity of Meyer, an engineering professor at the University of Auckland. Both he and Dow were also highly impressed by the personable and highly capable Peter Whittall, the man leading the project.[22]

Their appointments papered over Pike's stresses, but it would be much more difficult to airbrush away the costly and compounding setbacks that were already afflicting the mine's development.

FOUR
Trouble from the Start

There was little hint in Pike River Coal Ltd's 2007 prospectus that construction of the project's sole accessway to the coal seam was proving much more difficult and expensive than anticipated. By the time the prospectus was published in May 2007, the tunnel into the Paparoas had progressed almost 800 metres – about one-third of its full distance – and prospective investors would have been forgiven for assuming the remainder of the project would be almost as simple as coring through an apple. 'The tunnel is being constructed in hard rock type (gneiss) and is expected to be predominantly self-supporting over large sections,' the prospectus reported.[1]

In fact, there was every reason to assume the opposite was true – that the remainder of the 2.3-kilometre tunnel would be through difficult ground requiring engineering support every step of the way, and that costs would continue to balloon.

Les Tredinnick arrived at Pike in the winter of 2006 to work on the construction of the tunnel and ventilation shaft for contractor McConnell Dowell. At first he was one of four shift bosses, and later became site superintendent. The project was fortunate to have someone of Tredinnick's experience: he'd spent much of his career in the specialised world of tunnelling, including working as production superintendent for the second

Manapōuri tailrace tunnel, and driving the access tunnel for the Roa coal mine not far from Pike. Direct, intelligent and capable, Tredinnick was a man that big civil engineering contractors like McConnell Dowell kept on their contact list for when major underground contracts came up.

The tunnel project was running behind time even before the McConnell Dowell crew got started: because of difficult ground conditions and repeated washouts, construction of the last section of the road to the tunnel site had taken several months longer than expected, and the crew had to wait weeks to begin work. In the meantime, Tredinnick gave Harry Bell a hand with the mining of samples from the old adit at the top of the coalfield; this filled in time until he and his men could get access to the stone face from where they would begin boring into the mountain towards the coal seam.

They eventually got access to the site in August 2006. Like everything at Pike, their work environment was physically challenging and constrained. The area that had been selected as the portal, or entrance, to the tunnel was a large overhanging rock face draped with mature trees and thick mosses. From a small area pinched between the face and White Knight Stream they had to form a working platform, establish power, erect a 15-metre-high scaffold, and bolt and spray concrete on the face (a process known as shotcreting) to make for safe working conditions. When they started blasting the tunnel, they had to ensure that not so much as a single piece of stone was to end up in the stream, so they built enormous debris screens out of pine logs and slung them like curtains from wire ropes attached to rock bolts.

It was obvious from the moment that the first round of explosives was detonated in September 2006 that the ground conditions were terrible. Rather than the hard, self-supporting rock that had been anticipated, it was, as Tredinnick describes it, 'rotten' – broken, crumbly and wet.[2] The men drilled up into the roof and found that their rock bolts would not grab hold, and they had to insert steel rods and grout instead. They also had to manage what tunnellers call over-break – areas where oversized cavities opened up in the roof when explosives were fired.

Instead of being able to leave the rock largely unsupported apart from the occasional rock bolt, as they had expected, they had to painstakingly bolt, mesh and shotcrete virtually every metre. At the 190-metre mark the roof fell in, requiring careful remedial work. The incident was a warning sign that the ground conditions were not to be trifled with.

Conditions improved slightly through the next 700 metres, but progress continued to be much slower than expected. Pike had estimated that 90 percent of the tunnel would be built through rock that required little or no support, with only three percent in poor rock requiring intensive support. They had based that on work done by geologists Richard Cotton – the daring outdoorsman who had climbed down the western escarpment to collect coal samples for Terry Bates in the early 1980s – and Tim McMorran. The two men had spent part of the winter of 2004 clambering around in the icy bed of Pike Stream, examining faults and rock outcrops and making deductions about what the ground along the proposed tunnel alignment would be like to bore through. It was work that called for well-developed bushcraft and mountaineering skills as well as geological knowledge, as they scrambled across flood debris, truck-sized boulders and landslips, and bashed their way through dense forest nourished by the region's enormous rainfall.

As physically demanding as their task was, Cotton was unhappy that Pike was relying only on the pair's surface mapping of the ground, rather than also putting down exploratory drill holes. But the company wanted information 'quick time'; it was made clear to Cotton that drilling along the proposed tunnel alignment was not an option.[3] McConnell Dowell and its geotechnical advisers, URS New Zealand, also recommended that Pike sink four boreholes, at a cost of $500,000, along the tunnel route, but the work wasn't done.[4] There was an assumption within the Pike management team that the Department of Conservation, as landlord, would have disallowed it (although it was never asked).

Pike's decision to push ahead on the basis of Cotton and McMorran's surface mapping may not have been unreasonable, but given the risk that the tunnelling job could face costly surprises if ground conditions turned out to be worse than expected – tunnels being notoriously difficult to budget with precision, even in benign conditions – plenty of budgetary slack needed to be built in. At least, that's what Pike's consultant project manager Les McCracken thought. McCracken was engaged by Pike to oversee the tendering and contracting of the road, tunnel and ventilation shaft. Recognising the risk of surprises and cost blowouts, he built a ten to 15 percent contingency into his proposed budget. But, after a last-minute instruction from the Pike board to try and obtain a cheaper contract price than that provided by McConnell Dowell – the only compliant tenderer – Gordon Ward stripped out the contingency.[5]

There seemed to be a desire to reduce the apparent capital cost of the project. McCracken dubbed it a 'strategy of hope', and it cost Pike dearly: under the contract with McConnell Dowell, the construction company was paid according to the quality of rock and the rate of advance – the more difficult the rock, the higher the payment. As it turned out, nearly 80 percent of the tunnel was in poor rock requiring maximum bolting and shotcreting – almost the exact opposite of what had been predicted. Including an extension to the contract, which required McConnell Dowell to drill an area of utility tunnels known as 'pit bottom in stone', the project took twice as long as expected and came in more than 100 percent over budget.[6]

Nonetheless, Tredinnick's team did a fine job in challenging circumstances. Working 24-hour shifts, they crafted a smooth, horseshoe-shaped cavity measuring over five and a half metres wide and four and a half metres high, tweaking the levels where it was necessary to burrow deeper under stream beds, and driving successfully through the soft and broken ground surrounding the active Hāwera Fault. They were rewarded with a New Zealand Contractors Federation award for their efforts on such a large and difficult job.

On October 17, 2008, Pike announced that the project had finally hit coal. 'We have now very much de-risked the development and can begin operating as a coal mine,' Ward declared in a public statement. The timing was favourable. International benchmark contract prices for premium hard coking coal had risen 200 percent six months earlier, enabling Pike to settle its first coal sale contract at US$300 a tonne – three times better than the price forecast in the prospectus 18 months earlier.[7]

Reaching coal was an occasion of great excitement and news interest. One of those to witness the milestone was Grey District mayor Tony Kokshoorn, who had advocated so enthusiastically for the project. He was phoned in Greymouth by Peter Whittall and told the first coal would be reached that afternoon. 'I couldn't get up there quick enough,' Kokshoorn would later recall.[8] Local cameraman Patrick McBride filmed Whittall and Kokshoorn travelling in a specialised mine vehicle up the tunnel to where the McConnell Dowell men were meshing and bolting the roof. Whittall had carried with him from the surface a large lump of coal, about the size of a rugby ball. The tunnel project had been a long two-year haul, he noted, but this was 'a great day for Pike River Coal'. He held up to the camera a 'nice lump of Pike River coal … the first lump we've been able to get away from the face'. McBride affirmed the euphoric mood by setting the video to the score of U2's 'Beautiful Day'.[9] Kokshoorn called it a 'major milestone for the West Coast' that would have enormous economic spinoffs for the region. He regarded it as one of the happiest days of his life.[10]

But those more intimate with the progress of the mine thought Whittall had allowed himself a little too much licence in displaying the big lump of coal. There was certainly visible coal that day, but Tredinnick recalls it accounting for only about a third of the face. And Russell Smith, who had been hired in the first intake of Pike miners in June 2008, noticed from the photograph in the following day's newspaper that the solid lump held to the camera by Whittall bore no resemblance to the first coal handled by the men underground. That coal had been soft and crumbly. 'If you picked it up you could just crush it in your hands.'[11]

The 'first coal' story was entirely in keeping with the well-established tendency of Pike's leaders to err on the bright side. In a speech in August 2008, by which time both the budget and the timeline for the tunnel had been blown out by the terrible ground conditions, Gordon Ward waved off the setback with the comment that the project had 'not been without its moments'.[12] Making a new mine, he said, was comparable to running a race – 'not a short dash to the finishing line. Rather, it is very much in the nature of a multisports endurance event, like a business version of an Ironman or Coast to Coast.'[13] By the time full coal production was achieved from the middle of 2009, he said, the rewards would flow in the form of dividends for investors, gains for the environment, taxes and royalties for the government, and employment and wealth for the Grey district. 'You might say there are no losers!' he trilled.

Pike's progress was marked with an official opening ceremony at the mine site on November 27, 2008. A large marquee was erected and the event was attended by local dignitaries and politicians, senior executives from New Zealand Oil & Gas, and Peter Whittall's family. Only a handful of Pike employees were present. Miners were not invited. Minister of Energy and Resources Gerry Brownlee, the third most senior minister in the newly elected National-led Cabinet, was present to cut the ceremonial green ribbon. 'The successful development of this project required a carefully crafted marriage of good mining practice and environmental good management,' Brownlee intoned.

No one mentioned that, according to the 2007 prospectus, the mine should have already produced almost 250,000 tonnes of coal, but had not yet managed to get so much as a thimbleful away to the international coking market.

The following day, Pike River Coal Ltd's annual general meeting of shareholders was held at the mine site. In his speech to the assembled investors, Ward cloaked the operation in superlatives. 'I think we all have a sense that this is a special mine, in an unusual environment, and with a valuable product.' With the mine now 'on track to reach its steady-state production rate of one million tonnes per annum by mid 2009, we are increasingly being asked – what next?'

The answer to that question, Ward told shareholders, was that the company would soon start exploring the deep Paparoa coal seams, which lay under the Brunner seam that had only just been intersected by McConnell Dowell's tunnelling crew. 'Imprecise estimates' indicated there were a further eight million tonnes of hard coking coal that could be recovered from the Paparoa seam, and which would blend well with the Brunner seam coal, he advised.[14]

But to use Ward's own multisport analogy, the idea of exploring the Paparoa seams before getting the existing operation running smoothly was the equivalent of signing up for the Coast to Coast before learning how to paddle a kayak.

As well as hinting at bigger and brighter corporate adventures ahead, Ward declared: 'Our safety record has been excellent, with no serious harm incidents suffered by staff or contractors on the mine development during the year. In fact there have been none since first construction commenced on the access road three years ago.'[15]

Barely a fortnight before Ward made his uplifting speech, the men underground had in fact been confronted with a series of startling near misses, and Pike had run the risk of being shut down by the local mines inspector.

Les Tredinnick arrived at work one day in early November and read in the crew work diary that the men had experienced eruptions of blue flame from the tunnel face. There had already been an incident earlier in the year when a muck pile – a heap of excavated rock on the tunnel floor – had caught fire, fuelled by methane that had seeped into the rock from the coal seam. But what the workers were reporting this time was different.

Tredinnick began questioning the men about what had happened in his absence, and was told there had been two or three occasions already that day when gas seeping from the face had ignited. Then, just as he was being briefed on the issue, a rolling ball of flame rushed across the tunnel roof above his head. 'I was standing there, and then the next minute there was a whoosh. It was as if someone had thrown a cup of petrol up in front of me and ignited it. It was there, and then gone, in a flash.'[16]

On questioning workers from previous shifts, he discovered that at least ten ignitions had swept across the tunnel roof over the previous few days. The cause seemed to be the roadheader, one of Pike's brand new mining machines, designed to bore through rock and coal. The machine's steel picks were sparking against the hard, abrasive sandstone layer and igniting methane that had seeped from the rock.

Tredinnick was extremely disturbed. The flame that had swept over his head was frightening, and the presence of fire in a coal mine was a matter of grave concern. He immediately phoned his old colleague Harry Bell, with whom he had spent part of the previous winter gathering samples from the adit at the top of the Paparoas and who had worked under him as a shift boss for a few months over the summer of 2007/08.

Bell was shocked by what Tredinnick told him. Although by then he was semi-retired, his long career in mining, including his time as Greymouth's inspector of coal mines from 1978 to 1983 and chief inspector of coal mines in the early 1990s, had made him a stout defender of mine safety. His experience of hauling bodies from the Strongman mine following the massive gas explosion of 1967 had left him with a sombre respect for the potentially catastrophic combination of methane and flame.

During his few months working for McConnell Dowell on the tunnel, Bell had developed deep concerns about Pike's mine plan: a 2.3-kilometre single-entry tunnel, driven uphill through a major fault and into a gassy coal seam struck him as madness. He believed that to ensure an effective ventilation circuit there needed to be two drives through the fault, which was expected to produce high levels of methane. He also raised concerns about the volume of air being forced into the tunnel from a fan at the entrance: he considered it inadequate to dilute dust, fumes and gas to a safe level. He had even gone so far as to draw up an alternative ventilation plan.

On one occasion he told McConnell Dowell's construction manager Joe Edwards: 'If I was still the chief inspector of coal mines I'd shut the bloody thing now.' Indeed, if Pike's mine design had come across his desk when he'd been chief inspector – during an era when the regulator had considerable authority to rule on the technical and financial viability of proposed new coal mines – he would never have approved it. He had serious doubts about whether Pike understood the seriousness of the gas risks they were likely to encounter as they got into coal.[17]

To now hear from Tredinnick that men had been put at risk by at least ten separate ignitions of methane – and that this had occurred when the project had barely scratched the coal seam – was infuriating and extremely worrying. He wondered if any of his advice about the need to ensure adequate ventilation had been heeded.

Without hesitation, he rang Kevin Poynter, the Department of Labour's mines inspector in Greymouth, and told him of the ignitions. As it turned out, Poynter was already aware there had been a problem. Pike's statutory mine manager, Kobus Louw, had advised him there had been ignitions on two shifts. According to the underground coal mining regulations, Pike had a legal obligation to inform the inspectorate of such a serious event. Poynter had subsequently deemed Pike to be a 'gassy mine' under the regulations.[18]

Bell's information that there had been at least ten instances in the previous two weeks was news to Poynter, who, despite the watering down of the regulatory regime in 1992, had the power under the Health and Safety in Employment Act to issue a prohibition notice that would have prevented work until the cause of the ignitions was identified and fixed. Bell knew Poynter as a friend and colleague. Poynter had been in the coal mining industry for 30 years and had taken up the role of inspector only a few months earlier. But Bell also knew that much of the younger man's experience had been in opencast mines and at the non-gassy Denniston underground coal mines further to the north; he had little experience in gassy mines like Pike. He said to Poynter: 'Kevin, I'll give you a wee bit of advice. Stop them bloody mining until they fix the ventilation.'[19]

Only a week before Ward made his rousing speech to Pike's annual general meeting, Poynter wrote to Louw, demanding to know how many ignitions had occurred, and whether the company had considered the adequacy of its ventilation given that the mine was now in coal, and that the amount of methane would only increase as the project advanced.

'It is my opinion that the ignitions are probably caused by insufficient ventilation at the face,' Poynter stated. 'I have discussed this with the Department's Senior Hazards Inspector and we are extremely concerned that the issue of gas accumulation at the face and subsequent gas ignitions are dealt with urgently. While you have taken steps to minimise the exposure of the employees the issue needs to be resolved.'[20]

Poynter wrote again to Louw – coincidentally, on the day of Pike's annual general meeting at the mine site – asking for the documents relating to the ignitions. But there was no prohibition notice served, no prosecution and no mine closure. Poynter took the view that Pike was what it repeatedly told the public it was – a compliant 'best practice' company – and that the most productive way to deal with the ignitions was to negotiate a solution. However, he clearly harboured concerns that Pike had not reported the true number of ignitions, as it was legally obliged to do: he wrote again to Louw on Christmas Eve 2008, saying he had received formal notice of only two ignitions, but had been told by several people that there had been at least ten.[21]

Louw was furious that Bell had gone straight to the inspector; he felt this undermined his efforts to establish a culture among the men underground where all incidents and mishaps were notified through the mine's reporting system.[22] His response to Poynter's email was curt and dismissive – a reflection, perhaps, of the stress he was under as the mine ran further and further behind schedule.

'Don't know who fed you the information but there was a few ignitions on 4 shifts that I know of and that you should have the information (including the one at Hāwera Fault),' Louw wrote. 'If there is more then supervisors chose not to report hence I don't know of them and it is not been investigated [sic].'[23]

In the end, Poynter declared himself satisfied with Pike's solution to the problem, which was to stop using the roadheader and revert to the use of explosives to drive the tunnel forward. There was no enforcement action taken against the company.

A few weeks later, on February 13, 2009, Poynter closed the file. By then Pike was consumed with another crisis. The ventilation shaft, rising 111 metres from the coal seam to the steep, bush-clad ridge above, had collapsed.

The shaft, penetrating from the surface through the layers of island sandstone, mudstone and coal to the tunnel below, was a fundamental element of Pike's mine design. Fresh air would be drawn up the long 2.3-kilometre stone tunnel, and polluted mine air, carrying methane, dust and fumes, would be swept up the shaft. At the top of the shaft there would be a secondary ventilation fan and diesel generators; these would provide ventilation until the proposed main fan was installed underground. Long term, the surface fan would act as a back-up if the main fan stopped. A ladder with staged landings up the vertical shaft was also intended to suffice as an emergency exit for the underground workers until a more practical walk-out secondary egress was established further to the west.

There could be no large-scale mining of coal until the ventilation shaft was constructed. McConnell Dowell had the contract to build it.

At one stage Pike considered locating the shaft to the east of the Hāwera Fault, but on closer inspection of the proposed site – which was on a scree slope flanked by a razor-backed ridge – it became clear there would be significant stability problems in constructing it. A new site had to be quickly selected so preparatory work to drill the shaft could begin.

There was no site that was ideal, but the one that was finally chosen – to the west of the fault – had a couple of attributes: it had been previously drilled, so there was some knowledge about the geological conditions; and it was on a ridge between two steep gullies, with enough space for surface infrastructure. Nevertheless, there were risks: the top third of the shaft would pass through faulted ground, and the bottom section would be in soft coal measures – strata containing coal.

There were three possible methods for building the 4.2-metre diameter shaft. One was 'blind sinking'. Working from the surface, this would involve blasting out the shaft in stages, excavating the rock, and reinforcing the circular wall with bolts and mesh as the hole advanced. This was probably the safest choice, but because of Pike's mountainous location on Department of Conservation land, it was problematic. The department would have to be convinced to allow Pike to leave a pile of excavated rock on the tops, or a road would have to be built to the top of the shaft to allow the spoil to be trucked away, or the spoil would have to be removed from the site by helicopter. In any case, it would necessitate the construction of major infrastructure on the mountaintop, including ponds for treating water that would otherwise produce acid mine drainage. It simply wasn't feasible.

The second option was a method known as Alimak raise. A specialised platform with overhead rockfall protection would be brought into the tunnel to bore the shaft from underground, with the sides of the vertical hole being reinforced as it progressed. This had the disadvantage of being a fairly slow method, but both Les McCracken and McConnell Dowell's construction manager Joe Edwards argued that it was the most prudent, given the ground conditions. McConnell Dowell proposed in its tender submission to use the Alimak method.

The third method was to 'raise bore'. This would involve bringing a specialised drilling rig on to the shaft site and drilling a hole about 350 millimetres in diameter from the surface to the tunnel void below. An enormous round steel reamer head would then be brought into the tunnel and attached to a long steel rod lowered from the surface. The reamer would be drawn up to the top, boring out the full 4.2-metre hole. The rock spoil would fall to the tunnel floor, to be 'mucked out' by loaders and conveyed out of the mine. Once the hole was completed, the reamer head would be lowered back into the shaft base, removed and transported out of the tunnel. Only then could the sides of the hole be supported all the way from the top to the bottom with bolts and mesh by workers operating from a platform suspended from the top.

A fourth option, favoured by Louw, was to drop the idea of a vertical ventilation shaft altogether, and instead drive a second tunnel into the coal seam. He was concerned that ventilation was inadequate, and thought that a return tunnel would provide a far more efficient route for contaminated air to leave the mine. It would also serve as a second means of exit for the workers. Like Bell, he fundamentally disapproved of the single tunnel mine design.

Louw's idea was dismissed without debate. Pike management chose the raise bore method, theoretically the fastest option.

Months of preparation were required before drilling could start. The only access to the top of the shaft was by helicopter – there was no road or track to transport men and machines up the mountainside. A suite of helicopter services was needed, including two ex-Vietnam Iroquois, which operated from Taranaki in the North Island, a Taupo-based Russian Mil Mi-8 capable of hauling five-tonne loads, and several smaller West Coast-based choppers.

The first task was to strengthen the top 35 metres of ground by injecting, under high pressure, 95,000 litres of soupy grout into the fractured rock to knit it together. Next a platform had to be created, from which the raise bore rig would operate and where the ventilation fan would later be installed. To do this, diggers had to be disassembled, choppered to the site in pieces, and reassembled. The construction workers were then tasked with sculpting out a flat area about the size of a tennis court, and building retaining walls into the side of the mountain with concrete blocks freighted in by helicopter. The platform and lay-down area needed to be designed to take the load of the raise bore rig and all the equipment that would be needed for the operation.

The work crew toiled through the winter of 2008, living in igloos on the mountain for days at a time. The sun reached their tiny worksite for only a couple of hours a day, and permanent frost lay thick on the ground. Sometimes it snowed, and frequently there was torrential rain. There was no cell-phone reception; the only link with the outside world was via an often unreliable two-way radio and satellite phone. Kitchen and camp facilities had to be assembled, and then dismantled at the end. Toilets and showers were rotated out weekly by chopper.

Once the platform was completed, the raise bore rig – freighted in from Australia – was lifted up to the site. Helicopters flew back and forth for three days ferrying chunks of machinery weighing up to five tonnes, which then had to be pieced back together again.

Geotechnical advisers URS New Zealand had been commissioned by McConnell Dowell to design the system that would support the perimeter wall of the shaft after the raise bore rig had finished reaming out the 4.2-metre hole. But as the weeks passed, the firm's principal engineer on the job, Evan Giles, became increasingly worried about whether raise-boring such a big hole would work. The problem was that the ground around the 111-metre hole would have to support itself until it could be bolted and meshed down its entire length. Although it was outside the scope of his brief with McConnell Dowell, he and a young geotechnical engineer, Sarah Williams, reassessed the original core samples taken from boreholes previously drilled around the shaft site and analysed the ground conditions from scratch.

They ran the data through an internationally recognised analytical method called the Stacey McCracken framework. Named for South African Dick Stacey and Briton Allan McCracken, the method had been developed by the two geotechnical engineers in the mines of southern Africa to help assess how long the walls of vertical shafts were likely to stand unsupported before they started to collapse.

Giles and Williams concluded that the top section, which had been reinforced with grout, would be capable of standing unsupported for about a month. The bottom 17 metres, through soft coal measures, would stand for only about a week.[24]

Giles wrote a report setting out his findings, and the last of several meetings was held with Pike's senior executives, including Peter Whittall and Gordon Ward, on December 12, 2008 to assess the risks and discuss options. By then, however, the choices were limited. The raise bore rig was already sitting at the top of the mountain; every day it sat idle it was costing Pike $22,500. The pressure was on to get the job done.

There was passing discussion of the possibility of boring a small hole of about 1.4 metres in diameter, then working from the top to widen it to the full 4.2 metres, dropping the spoil to the tunnel floor and supporting the walls as the large hole advanced. This, though, would take months and the idea was quickly dismissed.

The more practical option was to reinforce the ground around the base of the shaft by installing 14-metre steel bolts drilled in from the roof of the tunnel, through the soft coal measures and into the hard island sandstone layer. This, too, would involve some delay as the bolts would have to be specially made.

Despite the concerns about the stability of the bottom section, Pike chose to take the risk on the raise bore method. And instead of bringing in the recommended 14-metre bolts to lock the lower section of ground together, it would use the eight-metre bolts it had on hand.

McConnell Dowell's Joe Edwards thought the risks of raise-boring could be managed far better if careful preparation were done to ensure that the work to support the shaft proceeded as rapidly as possible from top to bottom. In particular, he wanted Pike to lay concrete on the tunnel roadways, so the vehicles used to muck out the rock spoil could move quickly and not get bogged down in the soft ground. And he wanted a 'stub' – a working bay – constructed a short distance from the bottom of the shaft as a place where spoil could be dumped; this would reduce the time to clear each load away from the heap. Edwards thought that if every possible delay were stripped out, they would have a far better chance of getting the shaft supported right to the bottom before it started to unravel: 'It was all down to timing. Speed was of the essence.'[25]

But making the stub and concreting the roadways would have set back the start date of the raise-boring operation by a couple of weeks, and so the suggested preparatory work wasn't done. 'It would have delayed things, and so the answer was no,' Edwards recalls.

The raise bore rig started drawing up the reamer head on December 21, 2008. The small pilot hole had been drilled earlier in the month, and had almost immediately begun shedding small rocks. Before the big reamer head reached the top of the shaft 18 days later, bigger rocks started falling from the wall to the tunnel floor. Initially they fell in baseball-sized pieces; later they fell in hunks the size of fridges and vans.

Louw worked through the Christmas–New Year period while the raise bore crew were on site. It was a period of intense pressure, but he held on to the hope that the men would be able to install the necessary support structures down the entire length of the shaft before it began to irrevocably fail.

The slow grinding of the reamer head being drawn up through the earth was offset by the crashing and showering of falling rock.[26] Seth Tiddy, a young Cornish man employed as URS's site geologist for the tunnel and shaft project, was underground every day and could see and hear what was happening. He was taking photos of fallen rock, and emailing them to his bosses with the message: 'Forget the calculations, if rocks are falling down the small hole they're going to fall down the big hole.'

Once the hole was completed, Pike workers began mucking out the rock spoil from the pile at the bottom of the shaft, loading it on to a conveyor belt and transporting it out of the mine. The raise bore reamer had to be lowered back down to the bottom of the shaft and removed, now a far more complicated undertaking because of the rockfall. Les Tredinnick had come back from his Christmas holiday to find his painstakingly constructed tunnel and the large chamber at the bottom of the shaft a mess of mud and rock. It was now up to him and his men, in conjunction with the raise bore subcontractor, to remove the huge reamer head. They won Pike's approval to bring in a contractor with a thermolance to cut it off. It was tricky work; they had to operate from a high platform, and methane levels had to be carefully monitored because of the use of flame to slice through the steel cutting head. But the job was done without mishap.[27]

It was then a great rush to demobilise the raise bore rig (once again, involving multiple helicopter flights) and bring in the gear needed to support the shaft from the top down before it deteriorated further. Workers were winched down on a platform from the surface, and moved quickly to install rock bolts and mesh. Meanwhile they could hear huge rocks continuing to fall away from the shaft wall beneath them.

The men managed to support the top 70 metres, but before they could reach the lower section it broke away into a cavernous void more than 12 metres across. Trainee miner Scott Campbell was underground during the final phase of the shaft's collapse. He suddenly felt a huge shudder and could hear rocks falling; unable to assess the danger with only his cap light to illuminate the darkness, he leapt away while the debris crashed down. The front of the roadheader was damaged and couldn't be moved for several days. Campbell would later recall walking out of the tunnel afterwards and talking with Allan Dixon, who had been at pit bottom. 'He told me he'd felt the pressure change in the air when it happened.'[28]

The ground had failed in the coal measures above the eight metre rock bolts, creating a cavity that allowed the large blocks of island sandstone to collapse.[29]

It was simply too dangerous to go on. The bottom half of the shaft had to be abandoned. The only solution was to fill the lower section of the hole with concrete.

It took a thousand helicopter flights dropping cement down the hole to plug the failed section.[30] Bad weather repeatedly prevented the choppers from flying, and it was close to three weeks before the job was done.

The mine was now without its ventilation circuit. No mining could occur until the shaft was fixed. The decision was made – ironically – to use the Alimak raise method to create a dog-legged section that would bypass the collapse and reconnect with the intact section of the main shaft further up. It took until June 2009 for the 2.5-metre diameter Alimak to be completed. There were plans to install a second 2.5-metre bypass shaft to match the original air flow that would have been provided by the full 4.2-metre hole, but this work was never done.

Once again, the project had suffered a massive setback. Once again, its budget had taken a battering. There was an unsuccessful attempt to shove the blame for the shaft collapse on to URS and McConnell Dowell, and an insurance claim was lodged. That claim remained unsettled more than four years later.[31]

It was now more than two years since the 2007 IPO and the first shipment was running over a year late. There was still no coal coming out of the mine.

FIVE
Management Blues

The mishaps and misadventures of Pike River Coal Ltd were pushing the price tag on the project higher and higher, and shunting the prospect of revenue from coal sales further and further out into the future. In the 2005 paper by Gordon Ward and Peter Whittall that had established the business case for the project, mine development had been predicted to cost $124 million, with the first coal to be produced by September 2006. By the time of the 2007 IPO, $64 million had already been spent and the remainder of the project was expected to cost $143 million; coal was forecast to be coming out of the mine by March 2008. At the beginning of 2008, development costs were running at $196 million, with the first coal exports delayed until the end of that year. And by the start of 2009, the ventilation shaft collapse had once again pushed Pike's new mine-in-the-making further behind time: development costs were now $240 million and coal production had been delayed until later that year.[1]

As costs escalated and revenue remained non-existent, Pike had to repeatedly go cap in hand to its investors for the extra money needed to keep the project going.

In January 2008 – barely eight months after the IPO – the company raised a further $60 million from its investors, and borrowed $40 million from Liberty Harbor, a Goldman Sachs investment company in the United States. A condition of the Liberty Harbor deal was that Pike reach a 'steady state' rate of coal production by early 2009.

Ward promised investors that this fresh wad of money would pay for the remainder of the project and take the mine through to cash-generating production.[2] The project had been 'significantly de-risked in recent months'. While the tunnel was being driven in more fractured ground than had been expected, progress was improving nicely. Cost increases would be offset by the soaring international price of coal, Ward predicted. The subtext was that patient investors would be amply rewarded for their loyalty.

In April 2009, with the cavernous void in the lower reaches of the ventilation shaft now filled with concrete, Pike again had to go back to its investors, this time for $45 million. The cost of the tunnel and ventilation shaft alone had, by then, blown out from the tendered price of $23 million to almost $70 million. Pike advised investors that the rockfall had been 'unexpected by all parties including the contractor and its technical experts'.[3]

As ever, the optimistic message from Pike was that production of world-class coking coal was just around the corner: once the shaft was patched up, the mine would 'ramp up' to full production with its new hydro-mining system, and would ship its first coal to Japan by the last quarter of 2009.

New Zealand Oil & Gas, which had intended to sell its investment in Pike after the sharemarket listing of 2007 but was still on board as Pike's largest shareholder, was being repeatedly called on to shovel more money into the hole in the Paparoas, with Ward often springing requests for funding on the mother ship at short notice.[4]

In the long months of delay, while the tunnel ground its way forward through fractious rock and while the ventilation shaft crisis was unfolding, Pike's newly recruited miners were being paid to do very little. Russell Smith was one of the first to be hired by the company. Brought up in a small Southland mining community called Tinkertown, Smith had worked underground in the region's coalfields since the age of 16. In June 2008 he and his family had headed up to the West Coast to live, and he was hired by Pike as one of two experienced miners in the first intake of underground recruits. All the others were 'cleanskins', completely new to underground coal mining work.

Having been inducted, there was nothing much for Smith and the other men to do except watch the McConnell Dowell crew push forward on the slowly advancing tunnel. There was otherwise a good deal of hanging around and watching DVDs. Smith recalls being frustrated that there were few basic tools – hammers, nails, saws – with which to get on with ongoing maintenance tasks.[5] This period of well-remunerated underemployment went on for months.

At the same time, Pike's senior managers and technicians were under acute pressure to meet milestones, overcome the project's endless hurdles, and get the mine into production.

By early 2009, Pike was not only burning through cash; it was burning through people.

Kobus Louw, who had been recruited from the large South African energy company Sasol as Pike's tunnel and production manager and become, in late 2008, its statutory mine manager, wrote his short letter of resignation in January 2009. By then he had been at Pike less than two years.

Louw had started out as an apprentice at Sasol and gone on to complete a degree in mining. Over a period of eight years he had worked his way up the company hierarchy from miner to underground manager, eventually reaching the position of shaft manager overseeing a section of a large coal mine in the Secunda region of South Africa. He was only 31 when he was headhunted by Pike and brought his wife Petro and baby son Rinus with him to live in Greymouth. They came to love the peace and beauty of the place and were happy to be there.

Culturally, Louw was in familiar company. The project manager of McConnell Dowell's tunnel team, Corrie van Wyk, was South African, as was URS's senior geotechnical engineer Evan Giles. It was not unknown for Louw and van Wyk to break into Afrikaans during management meetings, much to the consternation of the New Zealanders and Australians.

However, by the time Louw arrived at Pike in March 2007 key decisions about Pike's mine design had already been made, and he was unhappy with several of them. It had already been decided, for instance, that the mine's main ventilation fan – a critical piece of infrastructure and central to the prevention of explosive accumulations of methane – was to be installed underground, rather than on the surface where it could be easily reached and re-started if it failed or an emergency occurred.

Also, Pike had already ordered three key items of mining machinery – two continuous miners and a roadheader – that were being built as prototypes by Australian company Waratah, a firm that had never made these machines before. Louw thought it was crazy for a greenfields project like Pike to go with anything other than tried and true equipment from proven suppliers. And he was surprised that Pike was reliant on a single-entry tunnel – he thought there should be two parallel drifts to ensure an adequate ventilation circuit, as well as provide a secondary means of exit from the mine.

He and the engineering manager, Tony Goodwin, an Australian who was a close confidante of Whittall's, had a fraught relationship. Goodwin made it known that he would not take instructions from Louw, and Louw felt that critical engineering decisions were being made in Goodwin's quarter without proper input from him, even though they would have an impact on the underground coal operation.

Louw's predictions about the prototype machinery turned out to be prescient. The roadheader, the first of the three Waratah machines to be delivered to site, didn't perform and caused constant frustration. When Louw tried to raise the problems with Goodwin's engineering department, the response was that the men were to blame for not operating the machine properly. The Waratah continuous miners – although idle during Louw's time at the mine – were, in due course, to be unmitigated failures.

For Louw, the debacle of the ventilation shaft was the final straw. His principal concern had been the stability of the shaft's top 35 metres,

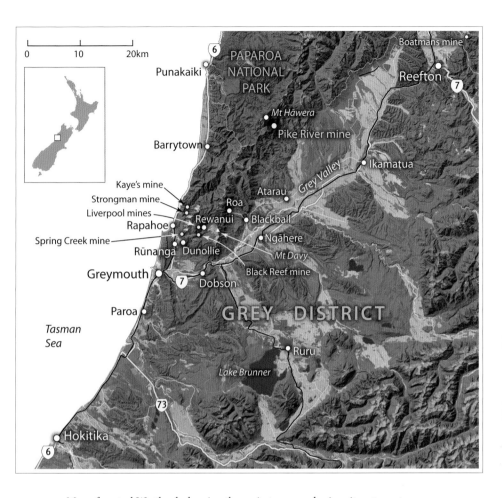

Map of central Westland, showing the main towns and mine sites. *Geographx*

ABOVE The town of Greymouth, population 10,000, straddles the Grey River and faces the Tasman Sea. It is overlooked by the Twelve Apostles Range. In winter a hard wind from the mountains – known simply as The Barber – often sweeps down the river, slices through the town, and leaves a thick white veil of mist across the Paparoas. *Stewart Nimmo Photography*

LEFT TOP Coal mining began on the West Coast in 1864 and unionism, imported by British miners, was strong. Miners' Hall, Rūnanga, seen here in 1910, was emblazoned with socialist slogans. A hundred years later the union representing mine workers on the West Coast was weak and marginalised. *Alexander Turnbull Library, Wellington, New Zealand (1/4-009396-F)*

LEFT Mourners at the graveside of men killed in an explosion at state-owned Strongman Mine in Rūnanga on January 19, 1967. Of the nineteen men who died, 15 bodies were recovered on the day of the explosion, and two more three weeks later. Two still lie in the mine. *History House Greymouth*

ABOVE The sheer west-facing escarpment at the edge of the Pike coalfield marks the boundary of Paparoa National Park. The Brunner and Paparoa coal seams are visible as black bands. *Stewart Nimmo Photography*

BELOW The adit in the Paparoa Range from which, in 2006, retired mines inspector Harry Bell was assigned by Pike River Coal to collect coal samples. The company planned to sell low sulphur and high sulphur coal, but Bell concluded separating the two would not be commercially viable. *MPR Collection*

ABOVE The road to Pike River Mine carved its way through native forest. In 2006, Peter Whittall (left) and Les McCracken observe progress made by the roading contractor, Ferguson Brothers. *Simon Baker/The New Zealand Herald*

BELOW The mine was ceremonially opened on November 27, 2008 with local dignitaries, politicians, New Zealand Oil & Gas executives, Peter Whittall's family and a few Pike employees present. Miners were not invited. From left: Gerry Brownlee, Gordon Ward, John Dow and Whittall. *Stewart Nimmo Photography*

ABOVE LEFT Kobus Louw, Pike's production and tunnel manager from May 2007 to October 2008, and mine manager to February 2009. He believed a second tunnel should be built, but management dismissed his idea without debate.

ABOVE RIGHT Doug White, Pike's operations manager from January to October 2010, when he became general manager; he was also mine manager from June 2010. A Scot, he had been deputy chief inspector of coal mines in Queensland and was regarded as a staunch defender of worker safety.

LEFT Neville Rockhouse, Pike's safety and training manager from 2006, who frequently expressed concerns about the ventilation shaft being designated a second exit. One of his sons was killed; another was lucky to survive.

Photographs: Stewart Nimmo Photography

ABOVE LEFT Tony Goodwin, an Australian electrician who was Pike's engineering manager from 2005 until early 2009. Loyal and hard-working, he was a close confidante of Peter Whittall.

ABOVE RIGHT Steve Ellis, Pike's production manager from the start of October 2010. Ellis also informally acted as mine manager for several weeks before the November 19 explosion, despite having not yet obtained the necessary qualifications for the role.

LEFT Masaoki Nishioka, an experienced engineer at Japan's Mitsui Mining, who came to Pike to commission the hydro-mining system in July 2010 and left three months later, fearing catastrophe. In 1993 he had noted that coal extracted from Pike boreholes contained so much methane that the gas bubbled out of samples.

Photographs: Stewart Nimmo Photography

ABOVE LEFT Tony Radford, chair of Pike River Coal's board of directors until April 2006, and a director until June 2011. As chair of New Zealand Oil & Gas, Radford actively promoted the Pike project.

ABOVE RIGHT David Salisbury, chief executive of New Zealand Oil & Gas from April 2007 to December 2011. Alarmed by the mine's mismanagement and constant demands for funds, he called in consultants Behre Dolbear Australia in early 2010. He was astonished when, in September, Pike chair John Dow announced that Peter Whittall would take over as chief executive. Salisbury believed both Whittall and his sacked predecessor Gordon Ward were responsible for Pike's recent mistakes and failures.

RIGHT TOP A McConnell Dowell worker engaged in construction of the 2.3-kilometre stone tunnel through the Paparoa Range to the coal seam in 2007. *Stewart Nimmo Photography*

RIGHT Early in November 2008 at least ten ignitions swept across the roof of the mine's tunnel. The cause seemed to be this roadheader – one of Pike's new mining machines designed to bore through rock and coal. The machine's steel picks were sparking against the hard, abrasive sandstone layer and igniting methane that had seeped from the rock. *Stewart Nimmo Photography*

ABOVE One of the two continuous mining machines purchased by Pike from Australia's Waratah Engineering. The machines were an unmitigated failure.

BELOW Pike River Coal miners with a Waratah continuous miner, July 2009. From left: Steve Torro, Daniel Rockhouse, Steve Cox, Mike Goudie and Dene Murphy. *MPR Collection*

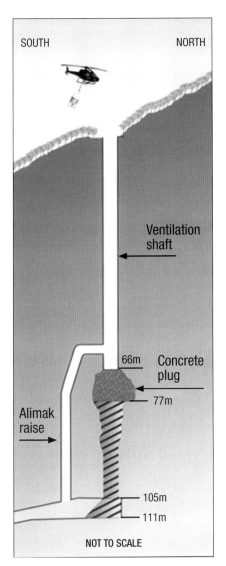

ABOVE LEFT Pike's ventilation shaft failed during construction in 2009. The lower section was plugged with concrete, and a smaller bypass shaft (Alimak raise) constructed. *Royal Commission on the Pike River Coal Mine Tragedy*

ABOVE RIGHT The 111-metre ladder leading from the mine up the ventilation shaft to the surface, which passed as Pike's emergency exit. A trial in late 2009 showed workers would find it impossible to climb up the shaft to safety in the event of an emergency. *MPR Collection*

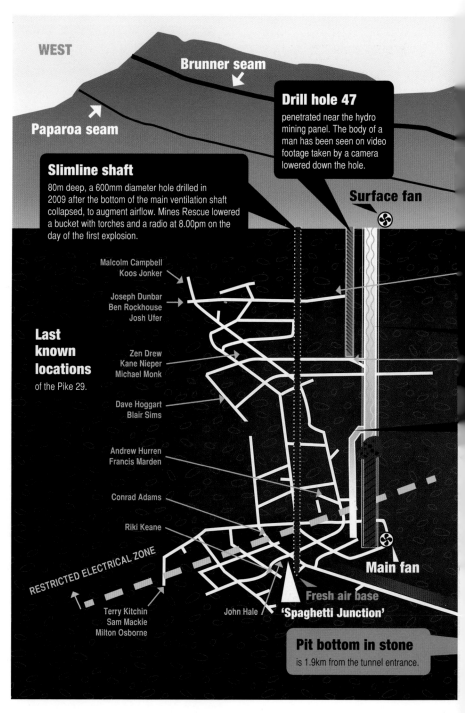

TOP Site of Pike River mine in the Brunner coal seam. ABOVE Underground layout of Pike River mine, including the last known locations of the 29 men who died.

PIKE RIVER MINE CROSS SECTION

EAST

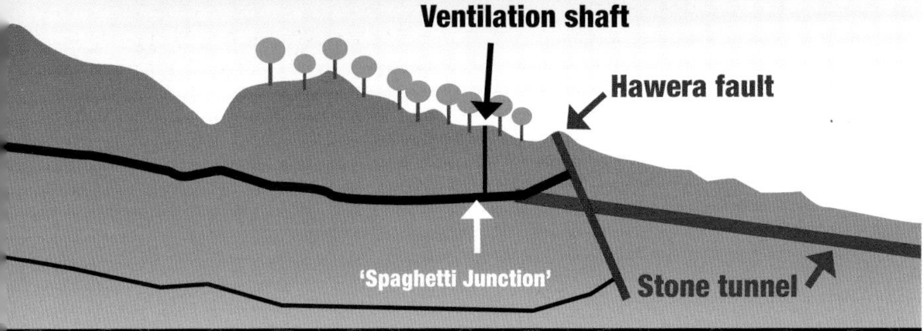

Ventilation shaft

Hawera fault

'Spaghetti Junction'

Stone tunnel

UNDERGROUND LAYOUT

Glenn Cruse
Chris Duggan
Dan Herk
Richard Holling
Willie Joynson
Stu Mudge
Brendon Palmer
Peter Rodger

Allan Dixon
Peter O'Neill
Keith Valli

Ventilation shaft
111m deep, connected the mine workings to the surface. The bottom section collapsed in February 2009 while it was being constructed. Some of the force of the explosions and toxic gases vented up the shaft. A steel ladder ran up the Alimak and ventilation shaft and was deemed the second means of egress.

Alimak raise
a smaller shaft constructed to by-pass the collapsed section of the main ventilation shaft.

Stone tunnel ('Drift')
is the ascending main access route into the mine, which ran 2.3kms from the portal to the coal seam. Part of the shockwave from the first 52 second explosion came down here to the portal. It provided the only access in and out of the mine.

Escape route
where Daniel Rockhouse was when the first blast occurred. He was at a diesel refuelling bay just off the main tunnel. He emerged with Russell Smith at 5.26 pm, an hour and 41 minutes after the 3.45pm explosion.

This diagram is not to scale and is based on interpretations of the evidence from the Royal Commission. The cross section axis is 20kms (approx) and the underground layout extends 4kms (approx). © MSO / MPR

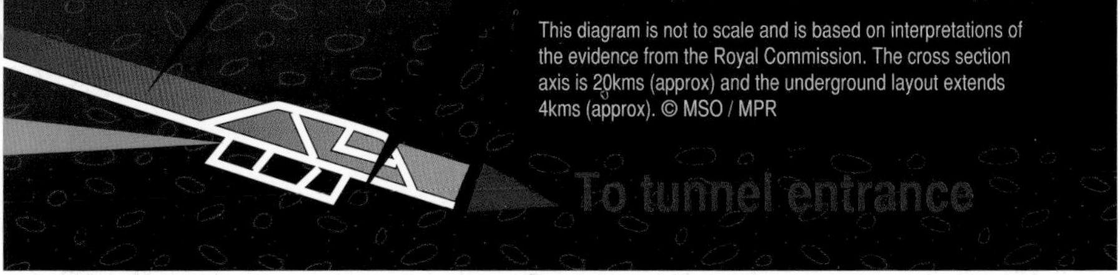

To tunnel entrance

ABOVE Hydro monitor in operation at Pike. The machine began use on September 19, 2010 but there were immediate problems, with methane levels spiking into the explosive range. *Stewart Nimmo Photography*

BELOW The guzzler, part of the hydro-mining system. The big steel wings gathered in coal being flushed from the face by the jet of water. The system was a prototype never used in a mine before, and proved unwieldy.
Stewart Nimmo Photography

ABOVE Pike workers operating one of the two Waratah continuous miners. The machines were one of the root causes of the mine's failure to produce coal as forecast, and a major reason for low morale among the workers. *MPR Collection*

BELOW One of the Juganaut loaders used to haul coal out of the mine. The machines were pushed beyond capacity and often broke down. Both Daniel Rockhouse and Russell Smith were about 1.5 kilometres inside the mine on Juganauts when the first explosion occurred. *MPR Collection*

ABOVE Pike's surface control room was equipped with modern systems, but when lights flickered, alarms sounded, red lights flashed on screens and the underground communication system went dead at 3.45 p.m. on Friday, November 19, 2010 emergency services were not called for 41 minutes. In this photograph, taken a year earlier, shift coordinator Marty Palmer monitors the mine's data feed. *MPR Collection*.

BELOW An incident form submitted by deputy Dene Murphy in June 2010. The mine's incident register was full of Murphy's angry and frustrated efforts to draw attention to failings in ventilation and gas management. *Royal Commission on the Pike River Coal Mine Tragedy*, Volume 2 and whether the grout would keep the ground solid. He had written a paper in October 2008 for the Pike River Coal board of directors asking for permission to extend the time allowed for the shaft work, so that the top section could be cored out using a small reamer head, with the hole then enlarged and supported in stages from the top. The board agreed. Shortly afterwards Louw took Petro and Rinus on a touring holiday around some of New Zealand's scenic highlights. In his absence, the December 12 workshop involving Whittall and Ward, URS's Giles and Don McFarlane, and McConnell Dowell's Joe Edwards took place.

On December 19, shortly after he returned to work, Louw was astonished to receive a handwritten instruction from Whittall that the raise bore be done in a single pass with the 4.2-metre reamer head. He went to Whittall and told him the decision would 'force me to look for another job as I did not agree with the company's business risk appetite,' Louw would later recall. If the raise bore operation failed, he predicted it would take at least six months to re-establish ventilation underground.[6]

Louw worked through the Christmas and New Year period of 2008 to 2009 while the raise bore crew were on site, clinging to hope that the engineering supports could be installed into the shaft wall before it failed. Deeply stressed and feeling that his professional input had been ignored, he resigned on January 12, having received a job offer from Sasol. 'Thank you for the opportunity to learn a great deal from this project,' he wrote in a brief resignation note to Whittall. He and his family returned to South Africa a few weeks later.

Louw was the first to pass through what became a revolving door of mine managers at Pike River Coal. Five more would hold the position between his departure and the explosion 22 months later.

By the time Louw left, Jonny McNee, too, had given up on Pike and quit. After two and a half years as the mine's geologist, McNee had come to the conclusion that the project would be a commercial failure because it would never produce coal to the specifications and volume Pike River Coal was promising its investors.

McNee had worked as a geologist in the West Coast coalfields since graduating from the University of Canterbury in the mid 1990s with a Master's degree in structural geology. His thesis was on the location and types of faulting and folding in the Greymouth coalfield; Jane Newman was his supervisor. It was a fascinating topic, but within a few months of being employed at the Strongman No. 2 mine he realised it was practically impossible to derive accurate structural models of the mine's localised geology from generalised observations and large-scale mapping.

The West Coast coalfields were formed when faulted slabs of crust were pulled apart and sediment filled the basins, just as icing fills the holes on top of a cake that has cracked during baking. In such a process, peat and sediment are first deeply buried and compressed, forming coal; later, tectonic forces squeeze the sediments up, effectively turning the basins inside out, thrusting the coal to the top of mountain ranges like the Paparoas.

Coal deposit, burial and uplift are all associated with complex and significant faulting. Detailed investigation, with closely spaced drilling and rigorous field mapping from the surface, are needed to develop workable geological models for a mine.

In the large state coal mines such as Liverpool and Strongman, the effects of geological complexity were often mitigated by being able to access coal from several different areas within the mine. For instance, if an unexpected geological feature stalled progress in one area, men could be deployed elsewhere in the mine and production would carry on. But in modern mines reliant on capital-intensive machinery and prescribed mining sequences – as Pike would be – the demand for production was such that even a temporary suspension would hit the bottom line.

From the outset, McNee felt that Pike's senior managers and consultants underestimated the importance of geology in mine planning. Whittall, engineering manager Tony Goodwin and consultant (and former Pike director) Graeme Duncan were all from Australia, where coal seams can extend more or less continuously for kilometres in vast tabular slabs with little complexity. A mine geologist could 'turn up every two or three months, collect the latest survey data and update the geological model'. On the West Coast, coal mining required a hands-on approach from geologists. There was a simplistic assumption among Pike River management that the Brunner seam at Pike comprised a thick continuous band of coal; in reality it was likely to be of variable thickness, intercepted by bands (referred to by geologists as splits) of worthless sandstone and areas where the seam could pinch away to nothing.

In conditions like these, the more drilling that is done in advance, the more confidently miners can predict what lies ahead. 'You can mine in a poorly defined area, but you have to have extremely flexible mining methods, or you have to accept you will run out of coal for significant periods of time.'[7]

With tunnel construction costing so much more than expected, it was difficult for McNee to get budget approval for the exploratory drill holes needed to help fill the yawning gaps in the geological database. Whittall, who signed off on even the most minor payments, watched spending like a hawk. (There was even an instruction to staff that colour photocopying was to be avoided.)

Initially, the requirement to get approval from the Department of Conservation was also an impediment to drilling; when McNee first started at Pike it took three months to get permission. But as trust built up between Pike's environmental manager Ivan Liddell and DOC's dedicated Pike liaison officer Mark Smith, approvals were processed within a few hours.

Pike was planning to make up for its sparse borehole data by drilling horizontally through the coal seam from inside the mine, to help figure out exactly where the coal lay. This in-seam drilling method would see drill rods probe hundreds of metres ahead, feeling out the roof and floor of the seam; it was a technique used widely in Australian coalfields. But McNee doubted that in-seam drilling would yield the same information as that which would come from densely spaced boreholes from the surface. For instance, the in-seam drill probes could easily miss a band of rock sitting in the middle of the seam, and they would not produce core samples that represented the variable sulphur profile of the Brunner seam.

Given that Pike's parent company, New Zealand Oil & Gas, had held the mining licence for a decade (and the exploration licence for a decade before that), it seemed to McNee that it was a bit late to be trying to accurately define the coal resource *after* the mine had been developed and hundreds of millions of dollars had been spent on infrastructure.

A further complication with in-seam drilling was that the long drill holes were likely to release large volumes of pressurised methane gas from the coal seam. This risk would have to be carefully managed.

Increasingly concerned about the quality of Pike's geological understanding, McNee sought out the advice of his mentor and former teacher Jane Newman who, 20 years earlier, had produced a model of the stratigraphy of the Pike coalfield. Newman suggested updating the model by incorporating the borehole data that had been accumulated in more recent years (not that there was a great deal of it). Whittall supported a budget application for the first phase of this work.

Just as in 2001, when she was helping Pike on the acid mine drainage issue, Newman was struck by how lightweight the geological data was as the basis for a mining operation.[8] Her view did little to ease McNee's worries.

There were aspects of Pike's coal that also added to McNee's doubts about the likely success of the project. The mine's promoters were correct in identifying the coal as having unique and unusual qualities, but he thought they were not necessarily attributes that would qualify it for premium prices, as the company claimed. Pike River Coal Ltd boasted that the mine would produce as much as 1.3 million tonnes a year of hard coking coal, with ultra low ash content of one percent. A significant problem that McNee foresaw with Pike's coal was that it was too low in coalification rank – a measure of the degree of heating and metamorphic change that the ancient peat mire has undergone during the burial process that turns it into coal. The higher the rank, the higher the percentage of fixed carbon in the coal; fixed carbon in coal is directly proportional to the fixed carbon in coke. High fixed carbon, high quality coke is valuable to steel manufacturers.

Because there is no such thing as a 'perfect' coal, coke-makers blend coals with different characteristics into mixtures that meet their steel manufacturing customers' requirements. Much as a baker might perfect the ideal loaf by taking a pinch of this and a tablespoon of that, a coking mixture typically contains coals from a range of mines and suppliers.

That being the case, McNee thought that Pike's coal might – just – fetch premium hard coking coal prices if it could deliver a product with the very low ash content that it promised. Ash is the remnant material left after burning, and is a mixture of minerals inherited from the parent plant matter, as well as clays and sands washed into the peat mire. Pike planned to lower the ash content of its coal by passing it through its brand new $20 million coal processing plant, which was designed to remove water and mineral content from the product. But there would be a trade-off between volume and quality: processing and washing the coal down to one percent ash would reduce the total volume of saleable coal. McNee's gut feeling was that for every five tonnes of coal to come out of the mine, only one tonne – at best – would achieve the ultra low ash specification.

Pike also planned to produce two grades of coal – one with low sulphur content of 1.2 percent and a lower-value coal with 1.9 percent sulphur. But McNee couldn't see how they could separate the two products using hydro mining, which flushes large volumes of coal from the face in a watery slurry.

In the end, there were too many reasons for McNee to leave Pike, and not enough to stay. Doubt about the project's viability, combined with high turnover within the ranks of middle management and a lack of respect for the role of geology all weighed against remaining. There were simply not enough technical resources to bring the project to fruition, and McNee knew the pressure would only get worse as the mine moved into production. He was also in the process of moving into a mining engineer's role at Pike, and he finally lost confidence in the place when he was told he would have to make the transition without formal training or access to further tertiary study. Instead, he was to receive 'on-the-job' training. He saw this as completely unacceptable – the position of mining engineer needed to be backed up by a recognised tertiary qualification.

When McNee quit in July 2008, work on Newman's stratigraphic model came to a standstill. Without McNee, there was no one to drive it forward. His replacement, South African geologist Jimmy Cory, who had not worked in a coal mine before, was keen to tap into Newman's expertise, and expressed strong interest in attending her training courses on coal geology. But it was difficult to find the time. 'The current workload just makes it impossible for me to leave the office,' he wrote to Newman in January 2009, a time when Pike was engulfed in crisis because the ventilation shaft was in the process of failing.[9] Cory never did manage to get to one of Newman's courses.

During his time at Pike, McNee had reported to the manager of the project's technical services department, the group responsible for mine design and long-term planning. In his two and a half years, he had three different bosses. New Zealand mining engineer Guy Boaz had come and gone from the role within little more than a year. Udo Renk, a highly qualified German mining engineer, had resigned after only 16 months, unhappy with the amount of geological exploration that had been done and concerned about the ventilation design. Renk had also been also in conflict with Peter Whittall, whom he considered a 'control freak' who often ignored advice and insisted on making every decision.[10] McNee's last boss was Kobus Louw, who filled in as technical services manager after Renk left, as well as fulfilling his main job as tunnel and production manager.

At the start of 2009, Whittall's loyal and hard-working engineering manager Tony Goodwin – one of the project's first recruits – also left. Several departmental managers had found Goodwin a difficult person to work with, but he and Whittall had collaborated closely on critical aspects of the project, including selecting the Waratah continuous miners and roadheader. Goodwin's influence was embossed on the project after his three-and-a-half-year tenure, but he was a man who kept much of the detail in his head. When he left to take up a more lucrative role back in Australia, much of that information went with him.[11]

Other members of the senior management team left too. Denise Weir, the widely respected and well-liked human resources manager who had fallen in love with the West Coast, returned to Australia in mid 2007. Her replacement, Mark Godwin, lasted less than a year.

Four years into the project and facing massive delays, Pike's management group remained in a state of constant flux. Instead of building up a tight and collaborative team of technical leaders to drive the project forward, Whittall and Ward were leading a group as fractured and unpredictable as the faulted ground in which they were aiming to mine. It wasn't unheard of for people to shout at each other in management meetings, with Goodwin one of the chief protagonists.

Ward almost always flew down from Wellington to attend the weekly management meetings, held on Wednesday mornings, although he was out of his depth when it came to geotechnical and mine design matters. His arrogant manner irritated many staff members: for instance, he would demand that Whittall's personal assistant Catriona Bayliss drop everything and make him coffee on demand, no matter how busy she was on other tasks.[12]

Whittall, meanwhile, was the undisputed boss of the project. Some, such as Bayliss, who started at Pike in mid 2008, admired his decisiveness, organisation and intelligence, and found him compassionate and thoughtful towards any office staff faced with personal difficulties.[13] Others, such as safety and training manager Neville Rockhouse, thought him a micro-manager who failed to support Rockhouse's efforts to implement the health and safety documentation he was creating.[14]

Denise Weir, while full of admiration for Whittall's sharp mind and enormous capacity for hard work, recognised his shortcomings; she wondered at times if it was necessary to try and modify his behaviour for the good of the project. 'Peter worked every hour that God sent. He had very strict ideas of what he wanted and sometimes people disagreed.

He is not an easy person, and sometimes he would flare up. He was fairly dictatorial, but I was always able to go to his office and speak to him. He is a man whose style you either liked or you didn't, but he was the boss. He probably was arrogant, but he had the weight of responsibility on his shoulders, and he was trying to pull together people from the four corners of the Earth.'[15]

Les McCracken, meanwhile, had long since distanced himself from Whittall. An experienced mining engineer, McCracken had been project manager for the Pike development since 2004 and served as engineer to the contracts for the road and tunnel construction. Initially, when Whittall arrived in 2005 to take up the role of general manager, the pair became good friends. McCracken, who commuted to Greymouth from his home in Ashburton most weeks, would regularly share dinner with Whittall, his wife Leanne and their three children; many evenings were whiled away discussing the project and sampling Whittall's admirable wine collection.

The relationship soured as McCracken became increasingly concerned about the wide gulf between Pike's public statements about the rate of progress and forecast coal production, and the repeated delays and cost blowouts that were occurring at the site. 'As a contractor, your reputation is only as good as your last job, and I felt increasingly uncomfortable about being associated with it,' he said later. He quit as project manager for the roading, tunnel and shaft contracts at the end of 2007, although he continued to act as the independent engineer certifying completion of contracts and approving payments to contractors.

McCracken made no secret of his concerns. On one occasion in early 2008 he upbraided Whittall following the release of a public statement predicting that coal would be produced by July that year. Whittall defended the statement, replying in an email: 'It doesn't say full production it says production. We are a coal mine and will mine coal. That is a true statement. We may even get black faces – make the South Africans homesick.'

McCracken replied: '[The] issue is an ethical one Peter. You know the assumptions that will be made by an average investor reading the statement and that is why the statements are being made. At best, that is skating on the edge of unethical behaviour.'[16]

The scolding from McCracken had little apparent effect. As the project lurched from setback to setback, the flow of sunny forecasts continued, troupes of politicians continued to bestow their praise on the mine that would set a new industry standard in modernity and excellence, and investors continued to reach into their pockets to keep the Pike dream alive.

SIX
Many Whistles Blowing

The first-time visitor to the Pike mine site could scarcely be unimpressed. From the brand new miners' bathhouse and $20 million coal processing plant to the 9.6-kilometre road and slurry pipeline that curved gracefully through the ancient forest, to the neat compound of offices and workshops that rested lightly in the wilderness – everything on the surface looked sophisticated, sensitive and modern.

Barry McIntosh found the place irresistible when he first visited in 2008. McIntosh had spent much of his working life underground in the coal mines of his home region of Southland. Like Harry Bell, he had started out as a teenager manhandling the 'empties' – the coal boxes that would be wheeled back underground to be filled by miners at the face. When he turned 20 he was able to put his name down to become a miner himself. He first became a 'floater' – working underground when an experienced miner was off sick, and eventually being taken on as a permanent 'cabled' miner when one of the older men retired. His first two years were spent driving roadways under the close scrutiny and instruction of an experienced Welsh miner; the next three years he was permitted to train in pillar extraction.

'If you wanted to survive underground, you listened. In mining, if you take a shortcut you don't get hurt, you get killed. Back then there was no paperwork, no risk assessments. The old guys would say, "This is how you do it" and you did it.' McIntosh learned the hazards of the industry the hard way: he lost a finger in Southland's Morley mine, and was twice buried in roof falls. Mine officials kept themselves alert to methane levels with the use of hand-held gas detectors, known as 'blinkies'.

McIntosh worked his way up through the industry, and in his late thirties reached the position of mine deputy. By the time he was approached by Pike to consider taking a job at the new mine he had been out of the industry for several years. He had bought a hotel business in Southland and then in Christchurch, but he had always loved the challenge of coal mining – listening to the coal, interpreting the smallest signs of movement in the seam, reading and assessing the risks. 'It's much like a house – it creaks and moves and it changes by the day and by the hour. You have your place where you're working, and you're constantly listening to how the place is working. You check it at night before you finish, and when you come in the next morning you know immediately if anything has changed.'

When he arrived at the Pike mine site and allowed his senses to take in the clear mountain streams, verdant bush and fleet of brand new mining machines – still gleaming white because they hadn't yet been underground – he said, 'Yep, I'll take it. I thought the place was unbelievable.'[1]

And Pike had indeed taken its obligations of environmental care seriously. No native podocarp larger than 60 centimetres in diameter had been chopped down without the explicit permission and supervision of the Department of Conservation, with the result that the project made only the tiniest dent on the vast expanse of the Paparoas. The company paid trappers to control pests over a 1,350-hectare area, eradicating hundreds of rats and dozens of stoats that would otherwise have preyed on the nests of endangered native birds such as whio and kiwi. The position of environmental manager, held by Ivan Liddell from early 2005, was one of the few posts within Pike's senior management not afflicted by constant turnover or Whittall's tendency to dominate and micro-manage.

At the end of 2008 the Department of Conservation bestowed an award on Pike in recognition of its efforts to marry mining with a high standard of environmental protection. And in early January 2009 the minister of conservation, Tim Groser, visited the site and lauded it as a 'showcase' of modern mining.

There was, of course, no mining going on at the time of Groser's visit because the ventilation shaft was in the process of collapsing and the place was in the grip of crisis. And even when the construction of the Alimak raise, bypassing the massive rockfall in the shaft, was completed five months later and ventilation restored, precious little mining could occur because critically important machines in Pike's gleaming fleet of modern equipment didn't work.

Peter Whittall and Tony Goodwin had overseen the decision in September 2006 to purchase two prototype continuous mining machines – 'miners' – from Waratah Engineering and had flown often across the Tasman to check on their construction. Why they contracted Waratah is a mystery – the company had never built continuous miners before. Pike's early planning documents showed the company had intended to buy from well-established brands such as Joy or Voest Alpine, and both companies had furnished Pike with tender documents for the provision of continuous miners.

The machines were critical to the mine's development. Equipped with large rotating heads and roof-bolting gear, their role was to cut roadways through the coal seam, creating access for the hydro-mining machine to be brought in for large-scale extraction. It was intended that the continuous miners would produce 20 percent of Pike's coal output and the hydro-mining machine 80 percent.

The two Waratah continuous miners, each costing $5 million, arrived on site in late 2008. They got off to an inauspicious start, with the brakes of one failing even before the machine was taken underground; it started rolling backwards down the steep road towards Pike Stream before coming to a halt.

The machines then sat idle for months while the ventilation shaft was fixed, and when they were finally pressed into active service they proved to be unmitigated failures. They stalled constantly. Their electric motors were positioned low to the ground and were not watertight, so whenever the machines were required to work through slush and puddles they would short-circuit. The hydraulic rams that lifted the rotating cutter head up and down against the coal seam repeatedly snapped, resulting in the head – which weighed about 12 tonnes – suddenly smashing down. The cutter head was wider than the shovels, so the shovels couldn't scoop up all the coal that had been cut. The tracks were poorly designed and would get bogged up with mud and muck. They were underpowered for the work they were required to do on Pike's uphill grade. The software that ran the electronics didn't work.

On one occasion a continuous miner sat broken down in the tunnel for a day, and men had to climb over the top of it to get to and from the coalface. The machines were seldom capable of operating for a full shift. Miners knocking off at the end of the day reported that the machines had managed to drive merely a metre or two through the coal. Sometimes there was no progress at all.

Complaints about the machines were repeatedly relayed to mine management. The message would come back down the line that it was the men who were at fault for not operating the equipment properly. Waratah Engineering, meanwhile, went broke in mid 2009 and was incapable of providing the necessary backup support. On one occasion the company's representative on the West Coast resorted to setting off on foot to the mine to work on the machines because his employer had not provided him with any funding to put fuel in his vehicle; he was endeavouring to make the point that the situation was untenable.

There were endless attempts to modify the machines in order to overcome their shortcomings, but nothing resolved the underlying problem – from front to back, they were a mess of ill-fitting features and unworkable design. In the eye of one experienced mining man, 'they looked like a four year old had sat down and designed a continuous miner.'[2]

Other machinery didn't perform much better. The roadheader, also from Waratah, was underpowered and broke down often. The hydraulics would jam if the smallest piece of debris – a bit of string or a rag lying on the floor of the mine – was sucked in with the coal.

Other machines also caused frustration and delays. The Juganaut loaders – used to haul coal from the face to the fluming station, from where it would be conveyed out of the mine in a watery slurry – were pushed far beyond their design capabilities by Pike's steep uphill grade. By the time they had been driven up the 2.3-kilometre access tunnel to the face they were 'puffed', Russell Smith would recall.[3] They were inclined to simply stop in their tracks, blocking the passage of other machinery.

Pike also had one old ram car that had already done extensive service in Australia. Although capable of hauling a much bigger volume of coal than the loaders, it broke down on virtually every shift. It was large, cumbersome, and often operated by inexperienced workers. Smith thought of it as a 'widow maker'.[4]

After ventilation was restored in June 2009, Pike's in-seam drilling contractor, Australian company Valley Longwall, was able to get to work. Because Pike had drilled so few exploratory boreholes from the surface, mine planning was heavily reliant on the long horizontal drill rods that would probe into the coal hundreds of metres ahead, curving upwards and down to locate the roof and floor of the seam, and establishing the angle of its ascent towards the western escarpment.

Gradually the drill rods built up a picture of the ground that lay immediately to the west of the ventilation shaft and the cavernous utility area known as 'pit bottom in coal', where pumping systems, sumps, electrical infrastructure, water storage and (in due course) the main ventilation fan were housed.

The picture was not good. Instead of the lustrous coking coal that Pike was promising to deliver to the market, a 200-metre zone of worthless shattered rock lay ahead. Right at the point where Pike expected to start mining coal, the seam simply vanished – it had been forced down between two faults. No one had known this feature, known in geological terms as a graben, was there. Pike's scanty exploratory surface drilling hadn't picked it up – it fell precisely between two widely spaced boreholes.⁵

Pike's public proclamations defied the gravity of these problems. On July 21, 2009, the company declared that the mine's production 'ramp-up' was underway, although 'teething issues' with machinery and the fact that some roadways had to be built through rock meant that the first shipment of coal would be delayed by a further six weeks – until November that year. Investors' nerves were calmed with reassuring commentary about the booming Chinese economy and news that China's demand for imported coking coal had reached record levels; coal prices were likely to increase the following year, the statement confidently observed.⁶

But away from the gloss of sharemarket announcements and press statements, morale underground was plummeting. Barry McIntosh, who had been asked by Whittall to take on a role in the mine's control room, a place kitted out with computer screens monitoring power, water pumps, communications and – eventually – the mine's fixed gas sensors, saw tensions rising between the men who were forced to work with failing machines and the engineering department that was trying to keep them going. 'A machine would break down and the fitter would come along and try to sort it out, but because it would be perhaps the third time on that shift that it had failed, the men would give the fitter heaps, and of course that would wind the fitters up. The attitude in the end became, "Well, they don't care, no one's coming in and really getting this problem sorted out, so why should we care?"'

At the same time, bottlenecks were building as the mine's workforce grew and contractors were engaged to install underground infrastructure such as pump stations, pipes and sumps. 'There was immense pressure, and everyone was getting their gear dumped around the control room waiting for it to be taken underground. It was a mess. That gradually flowed into an attitude of poor housekeeping underground.'⁷

Senior managers were meanwhile coming under acute pressure to get coal out. Nigel Slonker, Pike's statutory mine manager from April

2009 until his resignation just five months later, noted a cloud of financial uncertainty hanging over the place. Whittall spoke often about the Liberty Harbor convertible bonds, under which Pike was obliged to demonstrate that it was capable of reaching steady state production (66,700 tonnes a month, equating to 800,000 tonnes a year) by November 2009. Pike had already extended the target date once, and had been paying interest of 10.75 percent since April 2009.[8] If the November target wasn't met, Liberty Harbor could force immediate repayment of the debt or renegotiate the deal at a higher interest rate.

Faced with months of costly blasting and excavation through the stone graben before the coal seam could be accessed to the west, it was decided to put the continuous miners to work driving roadways in coal to the south of pit bottom in an effort to try and produce some coal.

Brent Mackinnon, an experienced mine deputy, arrived to take up a job at Pike in July 2009, just as the presence of the graben was being confirmed. Stockily built, plain-spoken and, like many of his mining peers, sporting heavily tattooed forearms, Mackinnon had learned his trade in the mines of the West Coast over almost a decade. Most recently he had worked at the long-established Roa mine, an operation run by a highly experienced and tight-knit crew of men working in deep Paparoa coal to the west of Pike.

Mackinnon was staggered by what he saw at Pike. For the first three weeks he was put through the company's induction process, much of which involved instruction in basic skills that he had mastered during his years in the industry. 'I thought it was a complete waste of time,' he recalls.

When he finally got up to the coalface in early August, 'No one was doing anything. Nothing was happening.' One continuous miner was underground but it had broken down, and none of Pike's four load haul dumpers were working either. Mackinnon wasn't given a crew to oversee and was forced to spend days 'wandering around doing nothing'.

One of his few supervisory tasks was leading a gang charged with bringing the second continuous miner underground. 'It was a debacle.

The tracks kept popping out, so we had to keep stopping and running the machine backwards so it could click back on to its clogs. The problem was that the track was smaller than it should have been, and so after a certain number of turns it would unmesh itself from its clogs. Then, when we finally got the machine underground, they wanted to test all its functions. Only about 70 percent worked – for instance, one of the drill rigs wouldn't deploy.'

Mackinnon couldn't understand why the machine hadn't been fixed and put through its paces before it was taken underground, but he got the message that senior management wanted it in the mine because 'some bigwigs were coming through'.

One evening in August 2009 a meeting of all mine deputies and underviewers was called to discuss Pike's ongoing problem of poor productivity. Whittall and engineering manager Nick Gribble – Tony Goodwin's successor – led the gathering, which was held off-site in Greymouth. Mackinnon recalls being told at the start of the session: 'Be open and honest.'

He took his bosses at their word. Indeed, he felt an obligation to speak his mind. 'I come from a place where you're encouraged to say your piece. As a supervisor it's your job to be the middleman between the men who have to deal with the crap [poor machinery] every day and the management. If you can't do that, what's the point?

'So I stood up and said, "Your machinery are pieces of shit. If they were dogs you'd take them out the back and shoot them."'

Mackinnon's honesty was not welcomed. 'Nick Gribble said he'd be ropeable if he heard anyone talking like that around the mine site. Peter Whittall didn't really say anything at all.'

The discussion drifted on inconclusively. Mackinnon sat through the rest of the meeting and then quit. 'It was obvious things weren't working, but they just weren't interested in listening.'

Aside from the hopelessly inadequate machinery, he had other concerns. He was unhappy with the large number of inexperienced workers underground, and he objected to the tendency of some of the underviewers to spend little time in the mine and to issue their instructions from the surface. And, like many of the other men, he worried that the 2.3-kilometre stone tunnel was the only means of exit from the mine: the ventilation shaft was tagged as the emergency egress but he knew it would be impossible to evacuate the workforce via the 111-metre vertical ladder.

Mackinnon had been at Pike only six weeks when he left to return to his old job at Roa. 'I just couldn't see a future in the place.'[9]

At around the same time Slonker – Pike's third mine manager in less than a year – also delivered a blunt assessment of the operation and resigned. Slonker had been managing a copper mine in Australia when he had been approached by Whittall to replace Kobus Louw early in 2009. When he flew to the West Coast for a mine visit a squadron of helicopters was shuttling backwards and forwards overhead, dropping 800 cubic metres of concrete down the collapsed ventilation shaft.

He decided to take the job, starting in April. To fill in between Louw's departure and Slonker's arrival, Whittall hired Australian Mick Bevan from a Newcastle-based mining agency, Palaris.

For the first two months of Slonker's tenure, work on the Alimak bypass continued, as did the drilling of a small 'slimline' shaft, intended as a stop-gap source of ventilation following the collapse of the main shaft.

Slonker was immediately under pressure from the top. Gordon Ward demanded daily text updates – even on weekends – advising how much progress had been achieved in the construction of the Alimak raise. At the same time South African geologist Pieter van Rooyen, who had moved with his young family from Namibia early that year to take up the post of technical services manager, was facing demands from Whittall to come up with a plan that would enable coal to be produced by September.

Slonker quickly became aware that Pike's promises to the sharemarket bore little resemblance to reality. Not only was underground roadway development excruciatingly slow, thanks to the hopeless machinery, but the area being developed to the south of pit bottom was so soft and broken that an enormous amount of roof support had to be installed, making for poor progress. Because Pike had encountered so many setbacks – in particular the failure of the ventilation shaft and the discovery of the stone graben – mine design was constantly changing. Even in the absence of such difficulties there would have been challenges in developing a mine plan suitable for hydro mining, which required roadways to be driven at a constant gradient to enable the watery slurry to flow away from the face.

There were problems with the mine's surface infrastructure too. As soon as the $20 million coal processing plant was turned on, it was revealed to be badly flawed. It had been designed on the assumption that it would receive the anticipated one million tonnes of coal per annum in a perfectly steady and continuous flow. Instead, small volumes were coming down the line in fits and starts, and the plant couldn't handle it.

By July 2009 Slonker had completely lost confidence in the project, and told Whittall so. 'I went to Peter and said, "Peter, this plan will not work. You are at least two years away from full production. These machines you've got are a waste of time and you need to order some new machines now." I said, "You need to order some Voest Alpine [continuous miners], and you'll probably have to wait for them for 12 months, but if you don't do it now you'll be another 12 months down the track and still faced with the same decision." The look on his face was very strange. I didn't know if he was going to hit me or cry. Then he said he didn't agree with me. He said, "These machines will do the job, these machines have been [specified] for the job."

'That was it for me.' He left on August 31, 2009.[10]

Slonker and Mackinnon were not the only people blowing the whistle on Pike's difficulties. Les McCracken had remained involved at Pike in a reduced role after quitting as project manager for the tunnel, shaft and roading contracts in late 2007. Back in 2007 he had thought relationships within the senior management team reporting to Whittall were 'toxic'. An established and well-connected contractor in the mining industry, he remained in touch with many of those who worked at, or contracted their services to, Pike. He also continued to hear news from his daughter, who worked in the company's environmental management team. And it appeared that by 2009 things were not improving.

McCracken was troubled by the personal dominance Whittall exerted over the project, and his tendency to reject advice he didn't agree with. From McCracken's years of involvement with complex projects, he knew that forceful, persuasive leaders such as Whittall sometimes created a climate where debate was stifled and those with contrary views were driven out.

He was also concerned about what he perceived to be a bullying style of management. 'People would be asked to go away and come up with a solution for something, and when they came back with their proposal and costings they would be told, "That's simply not good enough. Go away and do it again." Instead of building an environment where people work collaboratively to achieve the best outcome, you tend to end up with a whole lot of subordinates who will just give you what you want.'[11]

Through the winter of 2009 McCracken shared his concerns with another industry stalwart, Dave Stewart, the man who had been asked a decade earlier by Pike's consultant geologist Peter Gunn to come up with a pre-feasibility study for an underground mine, and who had trained many of the deputies at Pike. Both men were often in Greymouth on business and sometimes stayed at the Ashley Hotel at the same time. Their evening conversations repeatedly turned to the problems at Pike, in particular the low morale, the damaging effect that high management turnover was having on the men working underground, and dysfunctional relationships between the different nationalities on site.

The two discussed going directly to Whittall with their worries, but decided that would be pointless. By then McCracken had concluded that Whittall's management style was at the core of the problem. The Australasian Institute of Mining and Metallurgy's conference, the mining industry's premiere annual gathering, was to be held in Queenstown in late August that year, and they agreed that McCracken would use the occasion as an opportunity to speak directly to Pike's chair, John Dow.

McCracken and Dow had breakfast together on the final day of the conference. McCracken spoke frankly about his worries, and recalls saying of Whittall: 'The only way you will sort this out is to get rid of the guy.'[12]

Dow's recollection differs: he doesn't remember McCracken expressing such strong views about Whittall, with whom he had become good friends. '[Peter] was a very fine person, a very good engineer, and had a broad view of lots of things. We very much hit it off from the first time I met him. We were good mates.'[13]

Dow was aware McCracken and Whittall were not getting on, and he wondered at first what McCracken's motive was in raising concerns with him. Nevertheless, he knew McCracken to be a 'capable and consummate professional' and so took the issues about workforce morale seriously. He asked to meet with Dave Stewart at Christchurch Airport when both men were on their way home from the conference. Stewart confirmed everything McCracken had said. He told Dow there was a 'general unhappiness among the employees at Pike' and suggested someone with local knowledge and experience might be able to work alongside the mine officials and crews as a mentor, and make changes from within.[14]

Dow told Ward and Whittall of his conversations with the two men. 'I said to them both that I felt that Pike needed to take on board the advice of these guys, because they're senior and competent people, and then engage one of them if we could, or somebody with similar experience, to deal with the matter.'[15]

Dow passed Stewart's details on to Whittall and Ward, the very men overseeing the problems that had been disclosed to him, and left them to it. For six months nothing happened. It was not until February of 2010 that any concrete action was to emerge from McCracken and Stewart's extraordinary candour.

Pike continued to recruit miners in the expectation that production would get underway. Peter Sattler arrived to work at Pike in the middle of 2009. He flew from the balmy climes of Rockhampton to Christchurch, and travelled west by train across the Canterbury plains and foothills, through the scenic splendour of the Southern Alps, and around the shores of Lake Brunner to Greymouth. There was snow on the high country, and when he disembarked at Greymouth The Barber sliced through his Queensland shirtsleeves.

He had come to mine coal, but instead he spent the first six months at Pike excavating through the stone graben towards the coal seam. The rock was too hard for Pike's roadheader and Sattler recalls the machine's picks sparking against the stone.

As had happened months previously, when the McConnell Dowell crew were driving the last section of the stone tunnel, the pikes ignited methane gas seeping from the broken rock. Sattler regarded this as extremely serious. In any Australian mine where he had worked a gas ignition would prompt a full stand-down of men in the area, and a thorough investigation involving the mining inspector.

He immediately suspended operations and reported the event. He was shocked to come to work the next day to find that the night shift crew, which had come on after his crew had knocked off, had carried on working with the roadheader in the same place, as if nothing had happened. Sattler considered the decision to do this an appalling misjudgement, one that could have had catastrophic consequences. It was not until the following day that the mine management addressed the issue, ordering that the roadheader be withdrawn and the men instead use explosives to bore through the rock.

Many other aspects of the operation worried Sattler – in particular, the high proportion of cleanskins, the financial pressure on the company as a result of its failure to meet targets, and what he perceived to be slack regulation governing the underground coal mining industry in New Zealand.

Others held similar concerns. In August 2009 Sattler, along with three other experienced deputies who had come from Australia to take up jobs at Pike, completed a mining course entitled 'Demonstrating Knowledge of the Application of Regulatory Requirements to Manage an Extractive Site' at Greymouth's Tai Poutini Polytechnic. The men had been working at Pike for a few weeks by then, and Harry Bell was their course examiner. Bell completed formal one-to-one discussions with each of the men – a process known as a Professional Conversation – and also marked their written exam papers.

Bell was deeply disturbed by what the men told him. 'Their comments to me related to unsafe practices at the mine, issues with gas management, and the fact they had reported hazards which were not dealt with.' In Bell's experience, when a hazard was reported the standard practice was to display it on a noticeboard for all miners to see when they went to work, and to discuss it in daily 'toolbox talks'. The four men suggested to Bell that this wasn't happening at Pike, and that incident reports they filed were just being thrown in the rubbish bin.[16]

Bell's anxiety was heightened two months later when he received the men's written examination papers to mark. In response to one question, Sattler finished his extensive and detailed answer with this comment in block letters: 'PIKE RIVER HAS A SAFETY POLICY WHICH IT DOES NOT ENFORCE'. Another man, New Zealander Simon Donaldson, made a similar comment on his exam paper: 'I believe that West Coast coal mines generally have the paperwork side of the compliance with the Act and Regs under control. But this is not replicated U/G [underground] and more needs to be done enforcing this. Production pressures always seem to come first, which is wrong.'

Donaldson went on: 'Compliance with health and safety requirements sometimes takes a distant second place to production requirements. Coal mining is a hazardous industry that requires sound management practices to control all of the hazards. If we do not have safety systems in place and follow the Act and Regs we will make the same mistakes over and over, usually with deadly outcomes.'

In his exam paper, Australian miner Greg Fry also expressed deep concerns. 'Pike River is a single entry mine with no second means of egress. No code of practice or planned egress is in place. ... OSH [Occupational Safety and Health] inspectors should come to mines on regular intervals to identify hazards and bad practices before incidents.'

He continued: 'At Pike River there have been several unreported, unrecorded instances of frictional ignition and instances of multiple coal shots being fired simultaneously [banned in the mining regulations]. There should be a provision made in the legislation for a suitably competent officer to be able to be contacted by concerned employees and investigations instigated. Instances where untrained, unauthorised employees are operating machinery should be able to be investigated by a suitably competent appointed employee of that mine.'[17]

Bell marked the men's papers, and then picked up the phone to call Peter Whittall. It was many years since Bell had been chief inspector of coal mines, and he now had no direct involvement with Pike, but the matters raised by the deputies were too serious to ignore. He felt compelled to reassure himself that the mine's senior manager was aware of the issues.

The call went to Whittall's answering machine. Bell left a message asking Whittall to call him at ten the next morning at the Terrace underground mine in Reefton, which he was managing.

When the call duly came through, Bell had with him the Terrace's mine administrator, Kim McKay, as a witness. Bell wanted there to be no dispute about the conversation he was about to have with Whittall, who was also acting as Pike's statutory mine manager at the time: Nigel Slonker had quit and a replacement hadn't yet been recruited.

Bell told Whittall about the issues the men had raised – about health and safety policies not being enforced and about gas hazards that were not being dealt with. He also spoke of his long-standing concern that Pike's ventilation system was inadequate. 'I said, "I can listen to tittle tattle, but when it's down in writing I have to do something about it."'[18]

Whittall replied, 'Sometimes your officials let you down.'[19] He thanked Bell and said he would deal with the matters.

Having received the men's comments in his privileged role as an examiner, Bell did not want it to be known that he was the source of the information. He was also concerned that if he were revealed as the source, men might be reluctant to come to him in future with information and complaints. Whittall agreed to keep the conversation confidential.

Soon afterwards Bell heard from his nephew Allan Dixon, an experienced miner at Pike who was on light duties in the office as a result of an injury, that the promise of confidentiality had been broken and Bell had been revealed as the source of the information.

'For the greater good, so be it,' Bell thought to himself – the most important thing was that the mine's boss knew what was going on. He never found out what, if any, actions were taken by Whittall to address the matters he had reported.

There was frequent talk among the men underground that the mine lacked an emergency exit. If an incident were to block the 2.3-kilometre tunnel there would be no way out, other than up the ladder attached to the side of the 111-metre ventilation shaft.

Mining regulations stipulated 'suitable and sufficient outlets, providing means of entry and exit for every employee in the mine'. In early planning, Pike intended to seek approval from the mines inspector to use the ventilation shaft temporarily as an emergency exit, serviced by an electrical hoist capable of evacuating men out of the mine in half an hour. However, by 2005 this idea had been dropped and instead the shaft was to be fitted with a ladder.[20] The access agreement with the Department of Conservation allowed Pike to have four permanent emergency exits out to the western escarpment, but none had been developed.[21]

After the ventilation shaft collapsed and it was decided to fix the problem by putting in the smaller Alimak raise, bypassing the rockfall, Terry Moynihan, an experienced mining engineer, was contracted by Pike to oversee repairs to the ventilation shaft. Moynihan objected strongly when he was instructed to fabricate bits of the original ladder to fit into the Alimak section and to install the ladder without any landing stages where workers could pause for breath. He drew up some alternative designs; none were taken up.[22]

One of the few men to have climbed up the shaft was Matt Coll. An experienced mine project engineer, Coll had been contracted to Pike in early 2009 to install the hydro-mining system. He was physically strong, and served as a volunteer Mines Rescue Service brigadesman, undergoing regular and rigorous training to respond to mine emergencies. Yet even for someone with his physical fitness it was an exceptionally tough climb.

The first 35 metres up the Alimak shaft were straight up, then the ladder angled backwards for the next 18 metres – requiring the climber to lean back while making his way up the rungs. It took Coll seven minutes to get that far and he found it exhausting. There were then another 60 metres to climb – with a series of staged platforms – up the main shaft to the top. It was wet, slippery and dark. Coll realised it would be impossible to get out of the mine via this route if the air was not fit for respiration and breathing apparatus had to be worn.

Pike's safety and training manager Neville Rockhouse was also worried about the designation of the ventilation shaft as an exit; he asked the local manager of the Mines Rescue Service, Trevor Watts, to audit it. Watts produced a report in August 2009 concluding that escape via the shaft would be 'extremely difficult' in normal circumstances, and 'virtually impossible' in the event of a fire.[23]

Watts noted that any smoke and fumes in the mine would travel straight up the ventilation shaft, the very route that workers would be trying to use to escape. It would be like trying to escape a house fire by climbing out the chimney. And in any case the ladder could take only eight people at a time, yet there might be as many as 60 people underground on any given day. The report was passed on to Peter Whittall.

As part of a risk assessment of the emergency escape issue, a trial exit up the ventilation shaft was arranged in October 2009. Whittall was asked to attend, and Rockhouse, training manager Adrian Couchman and engineering manager Nick Gribble were also in the trial group. In the lead-up to the exercise there was some light-hearted email banter about which of the four men would make it to the top first.

When the day came for the trial, three of the four met at the bottom of the ventilation shaft. They waited for Whittall to arrive but he didn't show up. It turned out that Gordon Ward had more pressing corporate tasks for him to attend to: these detained him in his office.[24]

Couchman and Gribble decided to go first, clipping themselves into fall-arrest harnesses and making their way up. At the 50-metre mark they were completely exhausted. They looked at each other, said, 'We're not going any further', and decided to climb back down.[25] They were soaking wet and physically spent when they reached the bottom. Gribble said to Rockhouse, 'As long as my arse is pointing to the ground this will never be used as a second means of egress.'

It took until March 2010 to complete the risk assessment process. Meanwhile Rockhouse lobbied for the installation of a specially designed refuge chamber underground, at a cost of $300,000, as an interim measure until a functional emergency exit could be built. The spending was not approved. Instead, a rough shelter was made out of flame-resistant brattice cloth at the base of the slimline shaft.

There was a highly pressurised pipeline running right through the shelter, draining methane away from the coalface and up to the surface, but nevertheless the place came to be known as the 'fresh air base'. At November 19, 2010 it was the mine's only refuge. The ventilation shaft, despite being known to be entirely unsuitable as an escape route, remained Pike's only second means of egress. Planning for a proper emergency egress was underway in November 2010, but nothing had been built.

By the end of 2009 warnings were being expressed about Pike's equipment, workplace and management culture, and basic safety set-up. Whistles had been blown loudly and directly into the ears of the company's most senior executives, but the Pike story, as told through the company's official pronouncements, remained as bullish and upbeat as ever. Outsiders reading the company's quarterly and annual reports, or watching the PowerPoint presentations given by Ward and Whittall to gatherings of investors, would have struggled to discern the frailty of the operation.

In late October of that year Gordon Ward travelled to Hong Kong to speak to the Australian Stock Exchange's conference on small and mid-sized companies. He was there to tell them about Pike's 'showcase' project and its plan to mine the world's lowest-ash coking coal. He noted the 'upside potential' to mine deep into the Paparoa seams for a further eight million tonnes of hard coking coal, and made passing reference to the 'problem' with the ventilation shaft, which had since been 'corrected'. He spoke of how Pike's continuous miners were in operation, and of the preparations for hydro mining. He noted the ongoing buoyancy of international coal prices, and was hopeful of a lift in Pike's share price over the coming year.[26]

A few weeks later, he found the time to pen a lengthy submission to the New Zealand government's 2025 Taskforce. Led by the former National Party leader Don Brash, the group was charged with finding solutions to the country's flagging productivity. Ward waxed on about property prices, the New Zealand currency, and the need for more innovation and planning, and for a better climate for risk-takers and entrepreneurs. He cited the eminent economist Joseph Stiglitz, and derided the New Zealand 'number 8 wire' mentality. Once again, he made reference to Pike as a 'showcase of modern mining'.

By this time Pike had still not delivered so much as a teaspoonful of coal to the marketplace, and had managed to eke out and stockpile merely a few thousand tonnes of production. Not only was there a general mood of unhappiness underground, the company's major shareholder, New Zealand Oil & Gas, was deeply discontented too. Fed up with its endlessly disappointing offshoot, the company – still chaired by Tony Radford, who also continued to sit on the board of Pike River Coal – wanted out.[27]

SEVEN
Too Big to Fail

New Zealand Oil & Gas was stuck. In retrospect, Tony Radford must surely have regretted not selling off the Pike coal licence to an established mining company as soon as Gordon Ward obtained the necessary resource consents and the Department of Conservation access agreement back in 2004. Instead, the company had opted to support the development of Pike as a start-up underground coal mine and promote the public flotation of the company to thousands of sharemarket investors.

After the 2007 IPO, NZOG remained Pike's largest shareholder with 31 percent ownership, but thereafter it treated the mine company as a separate and independently run entity. Like a young adult moving out of home, Pike's corporate headquarters were relocated by Ward out of NZOG's head office and into its own Wellington office space, and the two companies ran separate administrative and financial systems.

Radford had made it quite plain that NZOG intended to sell its residual Pike shareholding within a year or two of the float, and that the coal mine was – in business-speak – a non-core activity. The company wanted to focus on its oil and gas activities, in particular its Tui oilfield, which had come into production in 2007 and was gushing cash into the company's coffers.

By 2010, however, Pike was performing not like the independent and capable young person that had moved out three years earlier, but like an unpredictable problem child who failed to communicate for months at a time and then turned up on the doorstep demanding the contents of the parent's wallet.

NZOG was looking for ways to cut the apron strings, but it was in too deep and there was no straightforward way out. It couldn't simply on-sell to another large investor, because any buyer who took on more than a 19.99 percent stake would be required under law to launch a takeover offer to all shareholders. It could, perhaps, sell down in chunks of ten or 15 percent at a time, but no sensible investor would want to risk ending up in the same situation that it was in – holding a major stake in an expensive and underperforming project, but with no control over how it was being managed.

Although Radford and fellow New Zealand Oil & Gas director Ray Meyer also sat on the Pike board, they were not there formally as NZOG's representatives and refrained from passing on any knowledge they had of the mine's operational and management struggles to the NZOG board. When the subject of Pike came up at NZOG board meetings, the pair would fall silent.[1] The remainder of the board and executive were in the same situation as other investors – reliant on Pike's upbeat statements to the sharemarket for information on how the mine was progressing.

Pike represented about one-third of NZOG's market capitalisation in early 2010; it was simply too big to let fail. Radford, the person with the longest continuous association with the Pike coalfield, was as cool and unfathomable as ever. He seldom made his thoughts about Pike's prognosis known to those around him, although his associates could see he was disappointed with the performance of his acolyte, Gordon Ward. It was also clear he agreed with the wider NZOG board that the company needed to dispose of its Pike shareholding. For NZOG's senior managers, trying to find a way through the Pike conundrum was endlessly frustrating and time-consuming.[2]

By early 2010 Pike had already gone back to its investors twice since the IPO, raising an additional $100 million in capital to pay for the project's soaring costs. It was always well supported by its loyal and patient shareholders, and NZOG also stumped up each time with its share of the extra money (although the company also clipped the ticket by positioning itself as joint underwriter and charging a fee for its services – Pike paid it over $300,000 for its underwriting role in the 2009 capital-raising).[3]

But Pike kept springing nasty surprises. One minute the news would be of good progress and on-target production; the next there would be a cash crisis, or an announcement that the first shipment of coal had been delayed yet again. In early August 2009, for example, Ward assured NZOG's chief executive David Salisbury that the project was on track to meet Pike's obligation to its US lender, Liberty Harbor, to produce coal by November that year. Just two weeks later, Ward told the sharemarket that the November date wouldn't be met and the first shipment would be in the first quarter of 2010, which meant Pike might have to come up with $40 million to repay the debt.

In September 2009 Pike looked as if it would need an extra $19 million in funding to keep the project going; by early 2010 that figure had been revised to $45 million at a minimum, *and* it had to repay Liberty Harbor.

In April 2010 Pike went back to the well yet again. It raised $50 million from its shareholders, and struck a deal with NZOG whereby that company took over the debt owed to Liberty Harbor (for which Pike would pay NZOG the handsome interest rate of ten percent). While all this was being organised, Pike was running out of cash and urgently needed to borrow $3.5 million from NZOG to pay that week's bills.[4]

It was a hard-fought deal. One of the conditions was that NZOG would have the right to buy a third of Pike's coal at a predetermined price for the entire 18-year life of the mine. NZOG demanded this extra sweetener in the hope it would help make its shareholding in Pike more attractive to other buyers. The company, as usual, took a juicy underwriting fee – this time of half a million dollars.[5]

Pike told investors this latest capital-raising would leave it with a healthy cash buffer of $18 million to absorb any further unexpected development costs. Once again, investors were promised that the mine would soon be producing a million tonnes of coal a year, and would generate riches of $4 billion over its lifetime.[6]

There was enthusiastic support for Pike's latest capital-raising from financial analysts. John Kidd of the broking firm McDouall Stuart, a joint underwriter, wrote excitedly on May 5, 2010: 'Another delay? Another capital-raising? We upgrade to Buy!'

Kidd advised that 'macro' drivers were in Pike's favour. International coking coal prices were soaring again and had doubled in 12 months. There was surging demand from China for steelmaking raw materials, and Australia's coking coal industry was struggling with logistical bottlenecks. Kidd went so far as to suggest that delays in bringing Pike into production had played to the company's advantage: if the mine had produced one million tonnes in the previous financial year, as it ought to have done, it would have earned a mere $180 million in revenue; instead, with coking coal prices reaching US$240 to $250 a tonne, it would earn $340 million.[7] Pike's catalogue of errors and misjudgements had, it seemed, been a lucky break for the company.

Such cheerleading helped keep the money rolling in, which kept the operation solvent, which kept Pike's cloak of grandiose ambitions from slipping to reveal a perilously frail organisation.

Even as Pike's supporters were writing out their cheques, Ward was calling on New Zealand Oil & Gas for further short-term funds to pay the bills: $6 million was needed to cover April and May outgoings.[8]

By the middle of 2010 NZOG had a total of $85 million invested in Pike shares and, in taking over Liberty Harbor's position as debt financier, had around $40 million (US$27.8 million) out on long-term loan to the company.[9]

Perhaps overlooking the role that NZOG's own penny-pinching on geological investigation in the early days of the project had played in

Pike's endless struggles, Salisbury and his fellow executives had arrived at the conclusion that the mine was being hopelessly mismanaged.

In the midst of the wrangling over Pike's latest clamour for capital, NZOG called in consultants to help it figure out exactly what was wrong. Behre Dolbear Australia was appointed to the task – ironically, the very same outfit that had written positively of Pike's merits three years earlier as part of the IPO prospectus.

In May 2010 Behre Dolbear furnished a two-part review documenting the flaws and failings at Pike. Signed by managing director John McIntyre, who had also authored the 2007 review, the review noted an 'air of despondency or resignation' at the mine site, and speculated that this might be because 'the equipment units are so poor that the efforts to get them to work efficiently are largely wasted'. Because of the high turnover of management, there was limited 'ownership' of the project. The consultants heard many people at the mine make comments about the problems that had been encountered.

Behre Dolbear's impression of the board and executive management was that their focus was more on the market – a reference, presumably, to the financial markets – than on the project, 'and there is a lot of effort being expended on presenting the project to the broking community'.

The report made some pointed recommendations: get rid of the hopeless Waratah continuous miners (it would be a 'miracle' if they ever performed as intended) and Juganaut loaders, and replace them with machines that worked. Review the design of the $20-million coal processing plant that, in its existing set-up, would probably never perform as it was supposed to. In an echo of former Pike geologist Jonny McNee's analysis, it suggested that processing Pike coal down to the one percent ash content that Pike had obsessively promised the market meant that yields would be low. It was probably better to sell the coal for a lower price with higher ash than spend money polishing it to premium ultra low ash standard.

Behre Dolbear also speculated that, given the much higher salaries being paid in Australia for experienced mining personnel, the Australians who had taken positions at Pike might be at the 'lower end of the scale in Australia or have had no clear path to senior management promotion' – an implied swipe at Whittall and the mine's recently appointed operations manager Doug White, as well as the numerous Australians working as undermanagers and deputies at Pike. It suggested the place might benefit from getting rid of current management and bringing in a mining contractor.

It observed that the geological complexity of the coal deposit was still largely undefined and, as a consequence, the mine plan remained in a state of flux. And it considered Pike's latest forecast – predicting full production by February 2011 – was 'optimistic': the project was still freighted with significant risk.[10]

Some of Pike's supportive investors may have thought twice about throwing good money after bad in the 2010 capital-raising if they had had the benefit of reading the Behre Dolbear report, but alas they did not. NZOG alone had the benefit of its insights.

The Behre Dolbear review gave New Zealand Oil & Gas the ammunition it needed to pressure the Pike board for improved performance, but it scarcely scratched the surface of the mine's true malaise. Indeed, NZOG's executives might well have saved their company the consultancy fee and learned more about the true state of affairs by simply spending time at the mine speaking to men like Dene Murphy and Alan Houlden.

Dene Murphy had started at Pike as a deputy in 2008. By 2010 the mine's incident register was littered with his increasingly frustrated and angry efforts to draw attention to failings in ventilation and gas management. A muscle-bound hulk of a man, Murphy went by the nickname Mad Dog on account of his sideline interest in BMX riding. He was no angel – he was disciplined at Pike on one occasion for taking leave from work without authority – but he was a careful, safety-conscious miner. He knew from bitter experience that failure to manage hazards underground could lead to catastrophic consequences. He had worked at the gassy Mt Davy mine, where two of his crewmates had been killed in 1998 in a massive outburst – a sudden eruption of gas and coal from the seam. He kept a photograph of his two dead workmates, Royden Stewart and Shaun Jennings, on his Greymouth living-room wall.

Murphy had got his first insight into Pike's management of serious incidents back in 2008, when he was working as a deputy on the tunnel construction. Ten gas ignitions had occurred before the hazard was addressed. Eighteen months on, the response to near misses and safety incidents hadn't improved much. There were few days when Murphy didn't file an incident report about something – insufficient ventilation, gas management, machinery problems, erratic communication between departments.

As a deputy, he was responsible for carrying out safety inspections on his shift. If he decided work should stop because of unsafe conditions, he expected his judgement to be respected. And he expected that when he reported a serious incident the matter would be properly investigated and acted upon. But at Pike this wasn't always the case.

One of the many incident forms he filed documented the failure of the hydraulic rams on the continuous miner. The rams sheared off and the 12-tonne cutter-head smashed down almost as soon as his crew started operating it. 'I was blamed as the deputy, and the miner who was operating the machine was blamed because he wasn't signed off to operate the machine. The rams were fixed but they failed another three times before it was proved to be an engineering design fault.'[11]

On another occasion Murphy was disciplined for making a judgement that he believed was in the interests of his crew members' safety. 'We were working in a roadway that was in shocking condition, really rutted and muddy, and the guys were getting knocked around. I decided to stop production and tidy it up. But I was had on by the underviewer on duty, who told me to get the guys back to cutting coal.'

The next day he was called to Peter Whittall's office. The human resources manager, Dick Knapp, was also present, and the door was closed. He was told by Whittall: 'If you are told you are to do something by your undermanager [underviewer], you do it.'

'I asked, "What if it contradicts what I have been taught and I believe the decision is wrong?" He just kept saying, "If you are instructed by the undermanager to do something you do it." All they wanted was for me to continue cutting, regardless of the conditions or problems.'[12]

The experience sent a worrying signal to Murphy. 'It really made me think that I had no standing as a deputy to make decisions. I thought, shit, what have we got ourselves into here if that's the attitude of the management?'[13]

Murphy knew he wasn't the only person who felt his judgement was being undermined. Back in 2008 he had worked with Kobus Louw, whose plans were often overridden. 'He didn't seem to know any more than us about why the changes were made and I think he was just as frustrated.'[14]

Louw and others left, but many men stayed because the Pike project was a route to financial security. 'A lot of guys had mortgages and kids, and Pike was going to be the answer to everyone's problems over here on the Coast. It was the promise of a better future for themselves and their families because it was such a massive coal resource. There was the hope that we could see out our time as miners on this project. I think everyone was in that frame of mind, that even though the conditions were hard they wanted to try and make it work.'[15]

But Murphy worried constantly about gas, and about the fact Pike had never appointed a manager whose sole duty was overseeing the mine's ventilation. In Queensland and New South Wales, mines were legally obliged to have a dedicated ventilation engineer, but in New Zealand the law was silent on the matter. Even though Pike's own internal documents stipulated that it would have a permanent ventilation engineer, such a person was never appointed. Instead, the role was subsumed into the wider functions of the statutory mine manager.

'Who is the mine ventilation engineer?' Murphy wrote in hard, block letters on one incident form pointing out inadequate ventilation in one area. 'Ventilation engineer required. … Require immediate feedback within four days – or I will write a formal letter to the mines inspector.'[16]

Along with many others, he worried in particular about an area known as the Thunder Dome – a cavernous space just to the north of Spaghetti Junction where the roof was eleven metres high. Because methane rises, and the enormous height of the cavity made it hard to ventilate, Murphy was concerned that the space was effectively a large gas reservoir. The problem was compounded by the effect of changes in barometric pressure: when one side of the mountain was in shade and the other side in sun, and if the fan happened to be down, the entire ventilation circuit could change within minutes. Instead of contaminated air travelling up the shaft and leaving the mine, it would be pulled back down and into the mine workings.

Murphy decorated many of his incident reports with foul language, for which he received a formal warning. 'I was told if I did that again I'd be down the road. But I was just so frustrated. I wanted the mine to work, but I wanted it to work for everyone. There were brick walls everywhere. You couldn't get feedback on any incidents. What should happen if there's an incident is that it should be reported to the oncoming shift, and there ought to be a detailed explanation of what happened, how to avoid it next time, and what to look out for.'[17]

Alan Houlden was another with serious safety concerns. He left at the end of April 2010 after only six months at the mine, fearful that the place was so poorly organised it could blow up.

Houlden had grown up in South Yorkshire, and gone to work at the deep Grimethorpe Colliery as a 16-year-old school leaver. His early career as a miner began on the surface, learning the ropes in the stockyard and coal processing plant, handling empty coal carts, learning where to stand to avoid being hit by falling material or becoming entrapped. When he first went underground, it was to learn the basics in a seam that wasn't being worked; he was taught about ventilation and why contraband items such as matches and cigarettes were outlawed. One of the first instructions he received was: '"Turn your light off. … Listen, what can you hear? Put your hand in front of you – what can you see?" You can't see anything.

Put it as close as you can, you can't see your hand. That's how important your lamp is. Before you come underground you check both your lamps are working. You check they are charged. That's the first thing.

'The next point was: "Have you got your rescuer on your belt? If you haven't got your rescuer on your belt, how are you going to get out?"'

His first underground job was manning a single button that controlled a conveyor. Then he spent time hauling supplies to the miners, before finally becoming a miner working at the coalface. Even after three years of training, a miner in his day was still not classed as experienced.

When Houlden went to work at Pike as a leading hand for sub-contractor McConnell Dowell, he found that young men considered themselves miners after just six months on the job. 'The young people didn't appreciate what was happening around them and how important ventilation is, and how important sticking to procedures is, warning people if there's something wrong… There was a culture problem in the mine. They had been stood down far too long, just waiting for the mine to get into production. When they finally got into production the machines they'd got were inadequate, so there was, yet again, more down time.'

Inexperience exhibited itself in frightening ways. 'One time we just couldn't breathe in the heading we were in. I brought the lads and the machines out, then went back down to find out what was wrong. The continuous miner wasn't working, and there was a young man just sat there at that heading and the [auxiliary] fan was off.

'I said, "How long has that fan been off? Have you turned it off? Have you informed the official it's been turned off?" So he says, "I don't know but I'll start it for you if you want." I said, "No, you're not starting anything. Just leave it alone until I get an official down here." And I went and got a deputy.'

The young worker appeared to have no idea that restarting an appliance in a gassy atmosphere had to be done according to strict protocol by a designated mine official because of the risk that it could create a spark.

There were incidents at Pike that suggested to Houlden that one part of the operation took little heed of the impact of their actions on other men underground. On one shift he noticed his personal gas detector flashing 'off the scale'. He stopped, withdrew the crew and worked with a deputy to find the source of the problem. It turned out that Valley Longwall's in-seam drilling crew had opened one of their boreholes to release the gas pressure on their drill rod: the hole was pumping methane directly into the general mine atmosphere. Because Houlden's men were uphill of the Valley Longwall rig, the gas had come straight up to where they were working. 'It was just another instance of the poor communication at the mine, and how stretched the deputies were.'

Often he would emerge at the end of a shift feeling ill and nursing a splitting headache as a result of carbon monoxide fumes from machinery, which the ventilation system was not adequately clearing away. On one occasion the auxiliary fan providing ventilation to the face where his men were was shut off without warning, and all the ventilation was redirected to where the continuous miner was working. 'Nobody told us. It just happened. ... It wasn't like working in a planned operation. It was as if everybody was their own little independent unit.'

Houlden went home one evening and told his wife, 'I'm taking that job I've been offered because that mine is going to go.'[18]

Ironically, by the time Houlden quit and Behre Dolbear had written its report there were signs that things were starting to look up at Pike. In February 2010, Pike's first ever shipment of coal left the Port of Lyttelton – albeit a mere 20,000 tonnes, less than one percent of the production that investors in the IPO had been promised. And in April the McConnell Dowell crew finished blasting their way through the stone graben and reached the main coal seam again.

Other big changes were afoot too. Peter Whittall, the man who had directed the mine's development since 2005, had transferred to Wellington, from where he continued to be Pike's general manager. The mine was still plagued by difficulties, was burning through cash, and was far from in steady-state production, but other business opportunities were coming Pike's way and the board of directors wanted Whittall to take a broader executive role in the company, including reviewing such opportunities.[19] Also Greymouth, with its limited educational choices, was becoming a little small for his family. Wellington had a university and better schooling opportunities.

Doug White was appointed to become the senior man at the mine site and arrived in January 2010 to take up the role of operations manager. White seemed to have all the right attributes for the job. He had trained as a young man in the Scottish underground coal mines, and worked his way up the ranks in Australia after emigrating there in the early 1990s. For several years he had run his own consultancy business in Queensland, providing training services to the mining industry, and running mines as a contractor. Before coming to the West Coast he was deputy chief inspector of coal mines in Queensland, where the lessons of historic mine catastrophes had been converted into rigorous regulation and tough policing of the underground coal industry. He also sat on the Queensland Coal Mining Safety and Health Advisory Council's Board of Examiners, responsible for statutory mining qualifications.

White was known to be a staunch defender of worker safety. He was seen by colleagues as the sort of man who might just as easily have been a union check inspector – a worker representative empowered by Queensland legislation to stop dangerous work – as a mine manager or inspector. But he faced a herculean task when he arrived at Pike. Morale was terrible. The men were still being driven to despair by the unworkable Waratah continuous miners. The middle managers responsible for getting the project on track were at their wits' end; they were the meat in the sandwich between the frustrated miners and the demands of company executives. Hundreds of incident forms had accumulated, awaiting investigation and action. A culture of blame pervaded the place, and the first reaction to problems was to finger a culprit rather than search for solutions.[20]

One of White's early tasks was negotiating with experienced local mining consultant Dave Stewart to audit the mine. The arrangement with Stewart was the upshot of the extraordinary high-level intervention almost six months earlier, when Pike's former project manager Les McCracken had pulled John Dow, the company's chair, aside at the conference in Queenstown and told him of his deep concerns about morale, working relationships and leadership at the mine.

While Dow had put Ward and Whittall in touch with Stewart and suggested they get him to help sort things out, matters had moved along at a glacial pace: it was January 2010 before Whittall and Stewart met. By that time White had arrived to work at Pike and Whittall was in the throes of transferring to Wellington.

Stewart, well aware of the deep unhappiness at the site, thought he could be of greatest assistance by spending time alongside the men underground and mentoring them in their roles. He knew many of them and had trained some of the deputies. He thought that by observing, listening and guiding he could help bring about change from within.

White, however, was less interested in mentoring and more interested in whether Pike was operating in line with the law. In an email to Stewart, copied to Whittall and mine manager Mick Lerch, setting out the main objectives of the audit, White said he wanted to make sure the mine was compliant with regulations, and that the statutory officials and electricians 'understand how to apply and maintain compliance'.

He then made an alarming statement that indicated he had deep concerns about standards and training at Pike: 'This is where I have the most difficulty as I find basic non-compliance every time I go below ground.'

White suggested Stewart spend time doing interviews to 'establish the level of compliance understanding that currently exists among our "officials". In doing things this way my objective would be that the "officials" realise the noncompliance's [sic] and organise to fix them instead of having to be spoon fed. ... I will shortly be addressing all of the current "officials" at Pike so they will be in no doubt where I am coming from, with your help on the compliance issues and reporting I intend to use more carrot than stick!'[21]

Stewart spent a total of 14 days at the mine between February and April, noting shortcomings and making recommendations, and emailing his reports to White, Whittall and Lerch. While Pike was talking publicly about ramping up to full production with hydro mining by July that year, Stewart's reports documented an operation that was still far from ready for large-scale coal extraction. The mine, he noted, lacked a rigid regime for regular monitoring of the underground atmosphere. When he arrived the fixed sensors required to transmit gas readings to the surface control room had not yet been installed; the one sensor fitted during his time at the mine was near the top of the ventilation shaft, poorly located for the task of continuously monitoring the air leaving working sections of the mine.

Roadways, he observed, were littered with debris and loose coal that presented a risk of spontaneous combustion. There was no programme of regular stone dusting, the essential practice of applying lime to the roof and walls of mines to render highly combustible coal dust inert. There was no emergency exit out of the mine; Stewart judged the purported second egress up the ventilation shaft to be 'impractical' for a large number of workers: 'only the fittest could do this route, particularly while wearing a self-rescuer'.[22]

Communication between shifts, and between deputies and underviewers, was patchy. There were too many inexperienced men underground and not enough experienced miners. And while there was a system in place for reporting incidents and hazards, there was no apparent procedure for converting the reports into actions. Miners and deputies told him they were putting in reports but hearing nothing back. They had begun to ask themselves why they should bother. Stewart encouraged them to 'keep on doing it, because there will be a response. It's just that they haven't obviously prioritised them at the moment.'[23]

In one of his reports to White, Whittall and Lerch, he warned: 'The effectiveness of any hazard management system is reflected in actions and feedback, otherwise the workforce will just see it as another company procedure that does not work.'

One of Stewart's major concerns was the quality of Pike's ventilation system. He was especially alarmed at the poor standard of the stoppings

– the barriers erected across certain roadways to direct ventilation, and ensure that clean air being propelled into the mine was not contaminated by dirty air being drawn out. Pike's stoppings were badly constructed and leaking. In some areas this meant contaminated air was re-entering the fresh air flow and being sent back up to the coalface where men were working. Stoppings were also regularly being damaged when explosives were used to drive roadways.

Stewart recommended that the mine's many temporary stoppings of brattice cloth and board be converted into permanent structures of timber and sprayed concrete to recognised standards. He had spoken about the problem to the mine's underviewers; they had told him there had been no training in the construction of stoppings because they didn't have the time. 'The underviewers I talked to were knowledgeable about these things but their focus was elsewhere,' he noted.[24]

The shortcomings identified by Stewart were extensive, but he offered practical suggestions, and felt that by the time he finished his audits in late April 2010 genuine improvements were in progress; he documented these in follow-up notes. Moreover, Doug White impressed him as having the experience, competence and commitment needed to turn Pike around.

White was indeed making a difference. Dene Murphy respected both him and the short-term production manager he had hired, Australian miner Bernie Lambley, although he got the impression White had his hands tied: he often looked frustrated.[25] Nevertheless, Murphy found him approachable and prepared to listen; his experience was that if he notified White of a problem, it would be dealt with promptly.[26]

Alan Houlden, too, saw that White was trying hard to make the mine work. 'He was pushing a lot more for standards, ventilation standards, support standards, general standards throughout the mine. You could see he'd got a mountain to climb but he was doing his best without interrupting production too much. ...

'I never heard an adverse word from [White] if you stood a job down for safety reasons. If you could justify to him why you had done it and it was a safety issue, no problem. The regime before he got there, to me, wasn't up to spec.'[27]

And within a few months at Pike, White had managed to do what no one else had hitherto achieved: he had wrung a grudging agreement out of Whittall that the Waratah continuous miners were at the root of the mine's production problems. By the time White arrived in January 2010 the men underground had been struggling for months to convince the company's senior management that the Waratahs were a hopeless failure.

None, though, had the persuasive power of Quintin Rawiri, whom White called on within his first few weeks at the mine to help him figure out what was wrong with the place.

Rawiri had started his career in mining as a school-leaver in his hometown, Huntly, and then headed to Australia after being laid off when State Coal Mines was restructured in 1987. Over the next 25 years he had developed expertise in coal and hard rock mines, tunnelling, mine recruitment, management and safety systems. In the late 1990s he and a small group of colleagues had set up their own company, Titan Mining, specialising in troubleshooting at underperforming mines.

White had been at Pike less than a month when he rang Rawiri and asked him to come to the West Coast to help figure out why Pike was failing. 'He said, "We're going shithouse, mate",' Rawiri recalls. '"The guys don't know what they're doing. Come and have a look at what we're doing and help us with our training issues."'

Rawiri flew to Christchurch early in February 2010 and drove over the Southern Alps to Pike. He spent three hours underground, speaking to the men and observing what was happening. That evening he went out for a meal in Greymouth with White and Whittall.

'We talked for a while and then Peter [Whittall] said, "So, what do you think of the operation? We've got pretty tough conditions." I said, "It's actually a pretty good operation." He said, "What about the men?" I said, "You've got good boys there, and they've been putting up with equipment that's not fit for purpose for a long time."

'I saw Doug look across the table at me as if to say, "Don't go there." Peter said, "What do you mean?" I said, "Your biggest issue is that you've got the wrong equipment."

'He then got quite cranky and talked about how they'd spent a lot of time speccing those machines, and how a lot of thought had gone into them and that they'd cost a lot of money to build. In the end I said, "Look, this is how it is: Doug has asked me to come over here and look at your operation. I can sit here and tell you what you want to hear and make you feel warm and fuzzy, or I can do what you pay me to do, which is to tell you exactly what I think. And I'm telling you the gear you've got is wrong, and you need to get an AMB20 in there."

'He said, "That'll never work in New Zealand. That'll never happen."

'I said, "If you're happy to keep doing what you're doing with your equipment getting one or two metres a shift, then carry on. It's not my business. I'm going to get on a plane back to Australia tomorrow." I just carried on eating my steak. The guy just would not listen.'[28]

Whenever White phoned Rawiri for more advice over the following weeks, the message was the same: get rid of the Waratah mining machines and bring in an ABM20 – a 100-tonne machine that could cut and bolt roadways in a single pass and was robust enough for Pike's steep grade. The machines were in critically short supply because of the enormous demand from Australia's booming mining sector, but Rawiri managed to locate one in New South Wales that was available for lease. In June 2010 White flew to Australia to inspect it. By this time Whittall had already been appraised of the Behre Dolbear report, which chimed with Rawiri's view that the Waratahs were central to the mine's failure to produce.

Whittall eventually yielded, telling White he could go ahead and lease the ABM20, but he'd be held accountable if it didn't work.[29]

The ABM20 arrived at Pike in August 2010, along with Lambley and three other men from Titan Mining who would help train the Pike operators. The Titan men drove the machine off the transporter and up to the coalface in a single shift; to Rawiri's knowledge, it was the first time in the history of Pike that a mining machine had completed the journey without breaking down somewhere along the way. In its first week the ABM20 cut almost 140 metres of roadway. In the year up to that point, Pike's two Waratah continuous miners and roadheader had achieved a combined total of not much more than 600 metres.

The air of despondency and resignation suddenly began to lift. It looked as if the fortunes of Pike River Coal might finally be on the turn.

New Zealand Oil & Gas, meanwhile, had continued to tighten its surveillance of Pike. Having taken over the Liberty Harbor bonds in May 2010 it was no longer reliant solely on Pike's ebullient statements to the sharemarket for its information: because it was now effectively Pike's banker it could demand much more detailed updates. From mid June 2010, NZOG executives met weekly with Ward and Whittall and heard details of operational progress, including the number of metres that had been advanced through the coal seam.

Pike's financial situation was still awful, even after raising the extra $50 million from its shareholders. At the time of the April 2010 capital-raising the company had forecast a cash surplus for the year to December of almost $18 million. But in July Ward revealed that by the start of 2011 the company would be $5.8 million short of cash – and that was on the optimistic assumption that the production forecast of 620,000 tonnes of coal would be achieved.[30]

Three weeks later Pike's financial controller, Angela Horne, gave NZOG notice that Pike was likely to soon be in breach of the terms of the $40 million bonds deal. It appeared almost inevitable that Pike would, for the fourth time in little more than three years, have to go back to the financial markets for yet more money.

NZOG was, however, conscious that some improvements were afoot at Pike. NZOG's executives had told Ward and Pike chair John Dow that they should stop penny-pinching and spend what was necessary to bring the mine into production, and they approved of the decision to bring in the ABM20.

NZOG chief executive David Salisbury was also encouraged that Dow was taking a more hands-on role at Pike – although the additional assistance Dow was providing came with a hefty price tag. In addition to his fee as chair of the board, Dow was contracted to Pike from June to December 2010 at $400 an hour (with a minimum retainer of $12,000 a month) to assist Whittall and Ward in their management functions.

Pike was also fortunate to have secured the services of a world authority on the hydro-mining method. Masaoki Nishioka – known as Oki – was held in warm regard by miners on the West Coast, where he had helped set up hydro mining at the old Strongman No. 2 mine and at Spring Creek over the previous two decades. He also knew something of the gassy nature of the Pike coalfield, having been involved in the drilling of the exploratory boreholes in the early 1990s and seen methane bubbling from the core samples.

At the invitation of Peter Whittall, Nishioka arrived at Pike on July 26 to assist with the final push to bring the mine into full production.

EIGHT
Marching to Calamity

Peter Whittall could scarcely have chosen a better-qualified person to come to Pike and oversee the commissioning of the long-awaited hydro-mining system. From early in his career as a mining engineer, Masaoki Nishioka had been involved in pioneering the system in his own country, Japan, as well as in Canada and New Zealand. He was, without doubt, a world authority on the method.

Hydro mining, which uses a high pressure jet of water to carve coal from a coalface, was found to be particularly well suited to the thick, steeply inclined and faulted coal seams of New Zealand's West Coast. It had been used as the main extraction method at the Strongman No. 2 mine from 1994, and at Spring Creek mine from 2004. Where other methods of mining would tend to leave large volumes of valuable coal behind in the geologically disturbed seams of the mountainous terrain, the hydro method could flush out virtually the full height of a coal seam.

Nishioka was familiar with Pike's plans to use hydro mining to extract 80 percent of its coal, and over the years he had answered numerous email requests for information from the company's former director and consultant Graeme Duncan. In 2008 he was asked, through his employer Seiko Mining, to draw up a plan for a hydro system at Pike and to put forward a price to supply the many items of equipment needed. However,

Pike had decided to purchase most of the equipment from a supplier in Australia (where hydro mining was virtually unknown), and Seiko was contracted to supply only the water gun to cut the coal from the face and the pipeline to carry the coal slurry away.

Nishioka had heard through the mining grapevine that things weren't going well at Pike, and was concerned that if the mine failed it would give the hydro method a bad name. If the system didn't work at Pike, he thought, it would be because of a failure of management, equipment or mine planning, rather than the technique itself.

As it turned out, an email from Whittall arrived in his inbox in mid June 2010, when he was working in Saudi Arabia on an oil plant installation. Whittall wanted to know if Nishioka were available to come to Pike to help in the 'ramp-up of operations' and become involved in a 'final critique of the installation, and also in the development and implementation of workforce training'.

When Nishioka arrived at Pike he was confronted with a situation that was much worse than he'd expected. There was a great rush on to finish developing the roadways that would give access to the first panel of coal to be extracted by the hydro system, as well as to install the water pipes, sumps and pumps required. But the mine's core infrastructure wasn't ready for hydro mining to begin. It was still reliant on the small backup fan located at the top of the ventilation shaft. The main fan, which would be the mine's principal source of ventilation, was not yet installed. And there was no emergency exit.

The hydro-mining equipment Pike had purchased from Australia was poorly designed. The pump that would feed water to the hydro monitor was unsuitable and had to be modified. The joints connecting the pipeline carrying high pressure water were prototypes; they leaked and appeared in danger of rupturing.

And the company had introduced unnecessary complexity into the system, making it more difficult for the miners to use. In particular, a machine called a guzzler was to be positioned about 18 metres behind the hydro monitor. Workers were to operate the monitor using levers located at the guzzler, which had enormous steel wings to capture the coal as it was flushed away from the face. This made the whole system clumsy and unwieldy to move whenever it came time to pull back to a new position to cut a fresh section of coal. The guzzler was yet another Pike prototype: it had not been used in a hydro-mining operation before.[1]

Nishioka was also unhappy about the area that Pike planned to start mining with the hydro system. The designated location was very close to pit bottom in coal, the section of the mine that would be a busy transit point for men and machines, and would house electrical, water and coal transportation infrastructure for the entire 18-year life of the mine.

It was also near the Hāwera Fault, an area high in methane gas that had migrated into cracks and fissures. Once the coal had been extracted, the goaf – the mined-out area – would be sealed off, but the residual coal and rider seam, the thin layer of high-ash coal above the main seam, would continue to emit methane. The goaf would therefore become a huge pocket of gas close to the nerve centre of the mine.

In Nishioka's view, sound and conservative mine planning would have seen roadways driven by continuous miners out to the western limit of the mine, with hydro extraction starting there and gradually retreating back towards the pit bottom area.

But Pike didn't have time to push out to the west and work back. It needed coal now. Under the original access agreement with the Department of Conservation, hydro mining was supposed to have begun in a location well to the west. The department had specified an area for 'trial' mining where it wanted to assess whether hydro mining would cause subsidence of the land above. But because Pike was plagued by endless delays and enormous cost overruns – the tunnel that took twice as long and cost twice as much as expected, the ventilation shaft that collapsed and had to be patched up, the thick stone graben that no one knew was there and took months to penetrate – it went to the Department of Conservation and asked for permission to start trialling the hydro system close to pit bottom. In the spirit of cooperation and goodwill that had developed between the department and the company, DOC agreed.

The first panel to be hydro-mined became known as the 'bridging' panel – so named because it would bridge Pike's financial gap by providing early income from coal sales.

By the time Nishioka arrived in July 2010, the choice of location, like the choice of equipment, was set. 'It was,' he would say, 'too late to rectify the mine plan.'[2] Nevertheless, he expressed his concerns about the set-up. He told Doug White he would not send men underground without adequate ventilation and a second means of exit, and he spoke strongly to Whittall of his concerns about ventilation and the design failings of the hydro system.

Nishioka was, by nature, courteous and polite. People who had known and worked with him for decades had never seen him raise his voice, or swear in frustration or anger. Slightly built and deferential, he may have felt somewhat apologetic about presenting the most senior people in the mine with such profound complaints. After all, he was effectively telling them their operation was poorly conceived, badly run and unsafe. In any event, his message was apparently not heard by either White or Whittall: both have denied he raised these concerns with them.[3]

The pressure to produce coal was palpable. On July 5, three weeks before Nishioka arrived, Pike chair John Dow had written to all employees offering a lucrative bonus payment if the hydro-mining system were up and running by September 3. If the target date were met and 630 metres of roadway development had been completed, each of them would get $13,000. The inducement payment would abate quickly for every week that the target date was missed: if start-up didn't occur until September 24, the bonus would fall to $10,000 (provided 790 metres of roadway development was completed). For every week of delay thereafter the payment would decrease by $2,500. The targets were 'readily achievable', Dow told employees.[4]

Assuming the workers earned the bonus and remained on the payroll, they would receive the money just before Christmas. It would be a handy boost to the festive season finances of men with families and mortgages.

The many contractors engaged by Pike to install underground infrastructure were not offered the bonus, however. Their exclusion was the cause of bitterness, given that many of them were small local operators and one-man-bands who were putting in huge hours at Pike to help the mine meet production targets.

As was his habit, Nishioka kept a detailed daily record of his activities. Through July and August he familiarised himself with Pike's wider production system, and made observations about activities in other parts of the mine. On August 6 he noted discussion at a mine planning meeting about the unresolved matter of the lack of an emergency exit from the mine.

On August 9 he wrote: 'Both faces are getting high gas content (0.4 percent) when [continuous miners] are not cutting. Need more ventilation air.' The following day he commented on the muddy, messy state of roadways, and noted 'many contractors are working underground. Some of them do not know the way to go out of the mine.' He noted: 'Smell of diesel fumes is bad. ... Main fan installation shall be commissioned as soon as possible.'[5]

Despite his reservations he pushed on, feeling unable to refuse to do the job he was contracted to do. 'I accepted adviser's work ... I have to do something. I cannot stay in the office sitting back on the chair and everybody was coming to me and [saying], "When can we produce coal?" I was getting, you know, that sort of pressure every day.'[6]

But he feared that Pike did not have a sound appreciation of the particular risks associated with hydro mining, and how those risks differed from other forms of mechanical mining. Continuous miners and roadheaders have a known production rate. Assuming the methane content per tonne of coal has been properly assessed, the amount of gas that will be released into the mine atmosphere can be calculated, and the amount of ventilation required to keep gas to a safe level can therefore be worked out. But hydro mining can cut through very large volumes of coal quickly, resulting in big surges of methane that need to be flushed away by the ventilation circuit and diluted to a safe concentration as they are carried out of the mine.

A large goaf mined out by a hydro monitor also presents the risk of a massive cave-in. Generally, the practice is to induce small controlled cave-ins as mining progresses. This reduces the risk of a major uncontrolled collapse that will generate windblast – a violent surge of methane, air and debris that can kill or injure workers and damage machinery and stoppings. Nishioka understood that Pike, which had to avoid surface subsidence under its agreement with DOC, was not planning to induce cave-ins; it expected the roof to remain intact. He regarded this as contrary to good mining practice.

Nishioka was by no means alone in his concerns about Pike's premature rush into hydro mining. The company was on the verge of introducing a major new coal extraction system that carried specific risks, yet a thorough process for assessing those risks and documenting how they would be managed had not been undertaken.

Australian consultant Bob Dixon of Palaris Mining came to the mine in late August and prepared a 'gap analysis', identifying tasks that still needed to be completed before hydro mining began. Dixon's list of missing or incomplete documentation included ventilation plans, procedures for dealing with 'plugs' of gas from the hydro-mining panel, plans for monitoring gas levels, plans for sealing off the panel once the coal was extracted, and risk assessments for key hazards such as windblast.[7]

The man responsible for advising Pike's insurers was unhappy too. On August 23, Jerry Wallace of Hawcroft Consulting emailed Whittall expressing concern about 'the lack of formal risk assessments one month out from the start-up of the first [hydro] monitor panel'. He thought it was 'unfortunate' that Pike was beginning hydro mining 'with many controls currently being developed but not yet implemented'.[8]

As the deadline for hydro mining approached, Nishioka's daily notes expressed mounting anxiety and documented myriad problems. On September 14, by which time the target date for the maximum $13,000 bonus had already passed, he wrote: 'Monitor extraction has to be started by Friday, at latest by Monday. ... Many works have to be done yet before starting monitoring.'

On September 15 the monitor and guzzler were hauled up into the mine, and on the afternoon of Sunday, September 19, the system briefly cut coal for the first time. The following day it was operated again, and almost immediately the methane content in the atmosphere shot up to five percent and kicked out power to the system. Two days later Nishioka ran the hydro monitor again, and again gas levels instantly rose above five percent. Nishioka wrote: 'It must be noted that it is safety hazard to continue the monitor extraction under this conditions [sic]. It is recommended that monitoring should be stopped until main fan becomes operational.'

On September 24 Nishioka wrote: 'After strenuous effort to produce 1,000 tonnes by the midnight of [Friday 24], which is the due date all employee are entitled to receive [NZ$10,000] bonus, several problems were highlighted.' Among those problems were that 'as soon as monitor started cutting coal, methane reading in the return airway came up over five percent. ... Reinforcement of ventilation shall be done before commencing monitor extraction. (Main fan was not yet operational).'[9]

Day after day, the problem continued. The hydro monitor would start working away at the coalface and methane levels would immediately spike into the explosive range. The only way Nishioka could keep gas levels down was by dialling back the water pressure and cutting less coal. At last, on October 1, Nishioka and the other members of the hydro-mining team agreed there would be no more use of the hydro monitor until the new main ventilation fan was installed.

A few days later Pike River Coal advised in a triumphant press release that the mine had passed its first hydro-mining milestone. The commissioning process 'has gone very well with all of the gear synchronizing, by and large, as expected'.[10]

It fell to Doug White to get the new main fan into operation. The scale of White's responsibilities had expanded considerably since he started working at the mine in early 2010. As well as continuing to fulfil his role as operations manager, he had picked up the job of statutory mine manager after the departure of Mick Lerch in June, making him the sixth person to hold the position in just 20 months. White also absorbed the role of mine ventilation engineer, with responsibility for the underground atmosphere, although he had no specialist qualifications in ventilation.

The new fan was to be the lungs of the mine, producing sufficient volumes of air to dilute methane down to non-hazardous levels and keep the atmosphere clear of dust and fumes. But, just as Nishioka was called upon to commission a system so badly designed it would probably never work as intended, White was the inheritor of a ventilation set-up that was deeply flawed.

As far back as 2006 Whittall and Goodwin had decided to purchase a main fan that would be installed underground, at the base of the ventilation shaft. When the idea was first mooted, the ventilation shaft was to be in rock on the east side of the Hāwera Fault. But over the next four years Pike's cascading series of mishaps and plan changes had fundamentally altered the risks involved in positioning the fan underground.

By October 2010, therefore, Doug White was commissioning a large electric fan that would sit in the gassy coal seam and be the centrepiece of a ventilation circuit that was inherently compromised. Despite the fundamental changes since the original concept was developed in 2006 and 2007, no formal process of reviewing the risks associated with locating the fan underground was ever done, even though the arrangement was probably unique in the world, and legislation in some places expressly prohibits the placement of a main fan underground.[11]

Despite its location in coal, the fan motor – along with several other items of electrical equipment, such as the water pumps – was not flameproof; it was set to automatically shut down if methane levels in the area rose above 0.25 percent. The back-up fan at the top of the ventilation shaft, which had served as the mine's ventilation source since mid 2009, would then kick in automatically.

The decision to place the fan underground alarmed experienced men. When Dene Murphy questioned it, he was told the fan motor was being positioned in a designated 'non-restricted' zone, an area where methane levels were to be kept ventilated down to 0.25 percent – effectively fresh air. In the 'restricted' zone, by contrast, equipment had to be flameproof or intrinsically safe so there was no risk it would create a spark that could ignite methane.

The non-restricted zone at Pike included not only the fan, but a large amount of electrical equipment and several variable speed drives, as well as the busy and congested Spaghetti Junction area through which ran a highly pressured methane drainage pipe, water and compressed air lines, and 11,000 volt cables.

The positioning of the fan in the non-restricted area made no sense to Murphy. 'I couldn't understand how it was a non-restricted zone when it was within ten metres of a temporary stopping into the main return, where all the gas was leaving the mine.'

Commissioning of the new fan did not go well. When it was put through its first test run on October 4, it sparked. Nishioka's notes for that day observed: 'Capacity of the [variable speed drive] is not enough for the main fan.' He met with Peter Whittall the same day, and spoke 'strongly' of his concerns about ventilation and the deficiencies of the hydro system.[12]

Two days later, on October 6, the back-up fan on the surface, still the mine's only source of ventilation, failed after a blade sheared off. 'All underground was gassed out,' Nishioka wrote. Power to the underground workings tripped out automatically, and all workers were evacuated.

The new fan was of no use: because of the high methane levels underground, it could not be restarted. The incident showed the faulty logic of placing the main fan underground: the very piece of equipment required to keep the mine atmosphere safe was unable to function when gas levels rose.

Commissioning of the main fan continued to be problematic; it would cut out two or three times a day, and trouble with the variable speed drives that controlled the motor continued.[13]

On October 12, both the surface fan and the underground fan tripped out.

Over the following eight days Nishioka's diary recorded a litany of problems. On October 11 Nishioka noted that the morning operations meeting had heard that the coal processing plant had achieved the target one percent ash, but at the cost of only 15 percent yield. On October 12 he described an apparent mismatch between the power supply to the mine and the voltage requirement of the variable speed drive operating one of the hydro monitor pumps: the pump started making a strange noise when it reached 2,760 revolutions per minute.

On October 13 the main fan tripped out again and the mine had to be evacuated.

On October 15 Nishioka noted that methane had spiked to 4.5 percent. On October 19 a 'large roof rock' had come down in the hydro panel. The main fan also tripped out – twice.

That day Nishioka emailed an industry colleague who was keen to hear news of Pike's progress. 'One monitor pump is running and started production, but running time is not so great, having water shortage, dirty water blocking up suction strainer, ventilation fan trouble, power outage, [coal processing plant] is unable to process coal (not enough capacity), and the like,' he wrote.

'It is much worse than I thought. I am now happy heading back to Tokyo tomorrow. I would think we should stay away from this project as it would not fly I am afraid.'[14]

Nishioka felt the only way Pike could get hydro mining operating successfully and safely was to start again: the set-up needed to be 're-engineered properly based on experience and knowledge of underground coal mining operation'. Virtually all the equipment would need to be replaced with machinery that had been proven to work in hydro-mining operations. And it needed to be run by people with experience in the system.[15]

On October 20 he left the mine. He had come to the end of his three-month arrangement with Pike. Rather than stay and persevere, he went: he feared the mine could explode at any time.

The hydro operation now fell under the direction of George Mason, an Australian miner. Mason had no background in the method. He had been appointed to the position of hydro coordinator on August 23 by White and Whittall, who preferred him over three local miners already employed by Pike who had experience of the system at Spring Creek.

Mason had spent several years out of the industry following an explosion at the Moura No. 2 mine in Queensland in 1994. Eleven men had died in the disaster; Mason had been undermanager on duty at the time. A year later he voluntarily surrendered his certificate of competency. He had also been an undermanager at the Moura No. 4 mine when it exploded in 1986, killing 12 men.

When Mason came to Pike in August 2010 he hadn't regained the necessary tickets to hold the position of underviewer. He was new to New Zealand, new to Pike, and new to hydro mining. He had been promised training at Pike, but apart from what he gleaned from Nishioka and the more experienced men under his command, he received no formal instruction in the method, which he found 'all very high tech'. Out of his depth, he tried without success to learn about hydro mining on the internet. Arrangements were made for him to visit Spring Creek to observe the system there but he never made the trip.[16]

Even so, Mason was comfortable with how things were proceeding at the mine. Nishioka had told him of his concerns about ventilation and high methane levels, but Mason didn't pass those concerns up the line to senior management or to others in the hydro-mining team. He thought the problems would be resolved with the installation of the new fan, and believed things were 'under control'.[17]

Soon after Nishioka left and before the main fan had been commissioned, the hydro monitor was pressed into constant service 24 hours a day, seven days a week. A decision was also made to widen the panel being mined by 50 percent; instead of spanning 30 metres, the roof of the goaf would now have to span 45 metres.

Miner Stephen Wylie was thrust into the position of deputy in charge of one of the hydro-mining crews in late October. He didn't ask for the role; he was simply told. He had a modest amount of experience in the method from having worked at Spring Creek, but was anxious to receive further training given that he would be in a supervisory role. Moreover, the men under him had no experience at all: a trainee on his crew had previously worked for a local builder who was contracted to do underground work at Pike, but he had never worked at the coalface. Wylie asked Mason about training, but was told he couldn't be spared from production duties. There was barely even time to exchange information with the other crew at shift changeover. 'It was always hurry, hurry, get your gear and get down the hole.'[18]

Despite moving to round-the-clock operations, Mason and the hydro crews were still failing to reach the required levels of production. The monitor was supposed to carve out huge volumes of coal and produce 80 percent of Pike's output, with the continuous mining machines expected to produce only 20 percent. Instead, the coal was 'coming off in very small particles as if we were sandpapering it off,' Wylie would recall.[19] Using the much-vaunted hydro system it had taken three days to mine one small area; he estimated it could have been done in an hour by the ABM20.

While the workers underground at Pike were battling to get the hydro system running, a parallel drama was playing out at Pike's head office in Wellington. On September 10, Gordon Ward, the former auditor who had driven the Pike project through its formative years under the tutelage of Tony Radford, and who, as its managing director, authored countless over-optimistic promises to the financial markets, was sacked by Pike's board of directors.

The board had been losing confidence in Ward since the middle of 2010; the mine was continuing to miss targets and they had had enough of his tendency to promise rates of development and production that were never achieved. A public statement announcing his departure was economical with the truth. The board thanked Ward for his 'significant contribution to the growth of the company over what has been an extended and often difficult period of mine development'. Ward, according to the statement, would finish up at the end of September; in reality he had been required to clear his desk immediately and been packed off on 'gardening leave'.[20]

The move against Ward met with the approval of New Zealand Oil & Gas. Having lost confidence in Pike's management, the company had appointed a special unit to work out a strategy to sell off its troublesome coal mining spin-off. Radford, who still chaired the NZOG board, shared the view that his loyal foot soldier deserved no more time at the helm of Pike.

But another announcement by Pike's chair John Dow, just four days after the news of Ward's departure, caught NZOG managing director David Salisbury off guard. Peter Whittall was to take over from Ward as Pike's top executive. Salisbury was astonished – Whittall and Ward were surely both responsible for Pike's chaotic recent history of mistakes and failures.

Nevertheless, if NZOG wanted to extricate itself from Pike with a decent return it had little choice but to keep quiet and continue to shovel in money so the mine could finally get into production. On September 23, NZOG learned that Pike was heading for a cash shortfall of $24 million. The mine's second shipment of coal – 20,000 tonnes that had been sent in early September to its shareholder and customer Gujarat NRE – hadn't met the promised ultra low ash specifications. Gujarat told Pike it would not take any more out-of-specification coal until at least the following year.

On September 28 NZOG agreed to tide Pike over with yet another short-term loan, this time for up to $25 million. By then Pike was in breach of the terms of its debt to both the Bank of New Zealand and to NZOG under the $40 million convertible bond deal, but both lenders agreed to waive the breach. Neither had any interest in destabilising Pike.

Three weeks later, on October 19 – the day before Nishioka left the mine fearing a catastrophe – Whittall announced to the financial markets that the Pike mine would produce only 320,000 to 360,000 tonnes of coal in the year to June 2011, about half the volume forecast three months earlier. By then Pike's board knew it would have to raise more money from the financial markets to fund the remaining development costs; it would be the fourth capital-raising since the IPO.

Unabashed by the disappointing production outlook, Whittall was full of confidence. He expressed his satisfaction with the commissioning of the hydro-mining system, which 'continues to progress well'. The outside observer could scarcely have guessed from the reassuring tone that the hydro workers were battling to keep gas levels down, that poor equipment was frustrating their efforts, and that production was a mere fraction of what had been expected.

The gulf between what Pike said was happening and what was actually going on was, as ever, vast. And not just in relation to the company's public proclamations to the financial markets – by late 2010 there was also a chasm between promise and reality when it came to management of the mine's key hazard, methane gas.

In its 2007 prospectus, Pike had minimised the risks presented by gas, claiming the Brunner coal seam contained only 'low to moderate' levels of methane. Yet data collected by its own consultants had showed that the Pike coalfield had what could best be described as moderate to high gas – more than ten cubic metres per tonne of coal at one borehole. Indeed, Whittall had spoken publicly of the need to reduce the gas content of some parts of the seam by pre-draining the methane before mining[21] – a practice involving drilling boreholes into the coal seam either from the surface or using an underground drill rig, and piping or venting the gas into the atmosphere, or capturing it as a source of energy. A lead time of many months could be necessary for the gas to bleed off to safe levels.

Pike was making use of in-seam drilling – but not for the purpose of draining the gas to safe levels; it was to make up for the inadequate number of exploratory drill holes from the surface and to help mine planners figure out where the coal seam lay. The methane that was released from the in-seam boreholes flowed under its own pressure through a four-inch pipe that ran downhill (against the natural inclination of methane to rise) via the mine roadways, through the congested Spaghetti Junction area, and then through a gas riser via the so-called 'fresh air base' and up the slimline shaft, the small hole drilled to augment ventilation after the collapse of the main shaft in February 2009.

By April 2010 the gas pipeline was under massive pressure. There was simply too much methane being forced through it from the in-seam boreholes, which burrowed hundreds of metres into the coal seam and branched off in multiple directions. Mine deputies repeatedly complained about the problem, and other workers were also worried. Les Tredinnick, the superintendent overseeing the McConnell Dowell crew who were continuing to do underground work for Pike, also complained about the drainage line: it was under so much pressure it could be heard hissing.

It was an email from underviewer Brian Wishart to the mine geologist, Jimmy Cory, which finally provoked action. Wishart wrote that violent surges of gas were coming through the pipeline, and warned that it could fail and send explosive concentrations of methane through the working areas of the mine, 'with plenty of Oxy not a nice scenario'.

He added: 'Just to bring to your attention the suspected findings of the American pit that recently exploded was centred around an inadequate methane drainage system.' This was a reference to the April 2010 explosion at the Upper Big Branch mine in West Virginia that had killed 29 of the 31 men at the site.

In-seam drilling should be stopped until a bigger pipeline was installed, Wishart told Cory. In a revealing insight into the way Pike's financial problems were perceived to be affecting decision-making, he added: 'I am well aware of the pressures we are under as a company but this should not be the pressure that possibly one day causes a serious incident. ... History has shown us in the mining industry that methane when given the [right] environment will show us no mercy.'[22]

Following Wishart's email a gas drainage specialist from Australia, Miles Brown, was contracted to advise Pike. Brown visited in April and again in May, and confirmed that the pipeline was too small and needed to be replaced with a 12-inch line. He also identified other problems. Pike had 'minimal data' on the gas content of the seam, and there was poor workforce knowledge of the risks surrounding gas drainage holes and pipelines. He urged Pike to act quickly over the coming months to gather data on the gassiness of the seam, and to assess the risk of an outburst, the violent ejection of coal and gas known to occur in highly gassy coal seams.

'In summary, Pike River has a chance to quickly understand what implications the inherent gas contents have on their production schedule. … The last outcome PRCL needs is a safety failure,' he wrote in a report to Pike on May 15, 2010.[23]

Brown wasn't the first to urge Pike to analyse and manage the risk of outburst. Back in 2000 Minarco had flagged it as a matter of concern, and in 2007 Behre Dolbear had said it needed to be 'on the management agenda'. And indeed Pike had drafted up an outburst management plan in July 2009, which stated: 'No mining will take place when the gas content of the coal is above the established outburst threshold level.' The sentiment was sound but the plan was never finished and no threshold was ever set stipulating how much gas was too much. Moreover, large chunks of the document appeared to have been simply lifted from Australia's Bulli mine without further analysis to make it specific to Pike.

Pike's technical services team took Brown's recommendations on board and began gathering gas data from in-seam boreholes. But by then the mine had squandered 18 months from the start of the in-seam drilling programme in December 2008, during which it could have built up a more detailed picture. As it was, the company had only scanty understanding of the gas content of the seam.

Money was available to upgrade the gas pipeline, too, but the work was never done. Instead, the stopgap decision was made – with Brown's support – to let the gas from some boreholes flow directly into the mine atmosphere, to be diluted and carried away by the ventilation system. 'Free venting', as it was called, succeeded in reducing the pressure on the pipeline in the short term, but it raised the background level of methane in the mine atmosphere.

When Brown came back to Pike in September, there was still no outburst management plan, free venting was continuing, and the upgraded pipeline was still in the planning. The push was on to start hydro mining; dealing with the gas issues had taken a lower priority than getting coal out.

Pike did have a detailed ventilation management plan. It ran to 78 pages and laid out the methods by which the ventilation system would be maintained and monitored to keep gas levels safe and workers free from the risk of asphyxiation and explosion. Among the key controls stipulated were the appointment of a dedicated ventilation officer, and the installation of twin systems to monitor the mine atmosphere from the surface. A network of real-time electronic sensors positioned through the mine would send data to computers in the control room. A tube bundle system – a set of tubes running through the mine – would pump samples to the surface for highly detailed analysis. This would continue operating even if underground power failed.

Pike largely ignored the plan. No dedicated ventilation officer was ever appointed. Among those anxious about the absence of a mine official focused on ventilation was technical services manager Pieter van Rooyen – he knew that such an appointment was a legal requirement in his native South Africa, as well as in Australia. Soon after his arrival to work at Pike in early 2009 he pressed Whittall about the matter, but was told that ventilation fell within the duties of the technical services department, and that New Zealand mining legislation didn't require the appointment of a ventilation officer.

Van Rooyen stressed that he had no experience in ventilation, and later took up the matter with Doug White, who agreed that someone should be trained for the role. Experienced underviewer Dean Jamieson was earmarked, but the training never occurred. Jamieson couldn't be spared from his other duties.[24] Instead, Doug White continued to absorb the role alongside his other duties.

Likewise, the promised belt-and-braces gas monitoring system wasn't installed. Doug White wanted to have a tube bundle system – regarded as standard kit in modern coal mines, and something that ought to have been installed as soon as the mine reached coal in November 2008 – and requested money be put in the budget for one. But then in September 2010 Pike had to go back to New Zealand Oil & Gas for more cash, budgets had to be redone, and the $800,000 to $1 million investment in the tube bundle system was put off until the next financial year. White, unhappy about the decision, began making arrangements to lease a system. His effort was stymied when Whittall advised the bank that was to provide the necessary finance that the tube bundle system was 'some way off'.[25]

And while Pike had made progress in installing real-time gas monitoring since Dave Stewart had commented on the absence of sensors in his audits, by late 2010 the system was effectively useless. Five sensors had been installed to measure the quality of air flowing over non-flameproof electrical equipment, such as the main fan motor; they were set to alarm at 0.25 percent methane but had a margin of error of plus or minus 0.25 percent and so were not fit for purpose.[26]

Three sensors were located in return roadways, measuring the quality of contaminated air being swept away from the coalface. Only two of them were set up to transmit information to the surface control room: of those two, one stopped working on September 4, 2010 and wasn't repaired, and the other stopped working on October 13, 2010.

The one remaining sensor measuring contaminated air hung from a piece of rope at the top of the ventilation shaft and was muddy, dirty and inaccurate. If methane levels reached the explosive level of five percent in the shaft, the sensor would record that as 2.96 percent. The sensor also got 'poisoned' when methane levels reached five percent, as occurred when the surface fan failed on October 6 and the mine gassed out.[27]

In the control room, which was elaborately decked out with computer screens, workers were adrift in a sea of poorly performing technology.

Barry McIntosh, the experienced Southland miner who had been entranced with Pike's pristine environment and modern equipment when he first arrived in 2008, was one of those deployed to the control room, but he and his colleagues were not trained in the monitoring system they were employed to oversee. No standard operating procedures setting out the response to gas alarms had been developed. By October 2010 a document had been drafted, but it relied on controls – such as a ventilation officer, an underground text messaging service, and a gas alarm logbook – that either didn't exist or hadn't yet been put into effect.

The controllers weren't even told where the gas sensors were located. When the hydro monitor was being commissioned and was regularly sending methane levels soaring, McIntosh would keep a vigilant eye on the sensor that he believed was positioned close to the newly installed main fan – an assumption he made because his computer screen showed an image of the fan, and listed critical data such as gas readings alongside.

As soon as gas levels rose he'd shut down the monitor from the control room and advise the miners through the underground communications system that they were 'putting too much methane out'. But he had no idea that the sensor he was watching so closely was not located at the fan at all. It was dangling by a piece of rope near the top of the ventilation shaft. By the time contaminated air passed over that sensor it was significantly diluted. If the sensor registered, for instance, 1.67 percent, methane levels at the coalface would be in the explosive range.

The deputies and underviewers working underground carried handheld gas detectors and recorded the results of their testing in daily reports. However, there were not enough detectors to go around: Nishioka, for instance, often couldn't find one to take underground. Control room operator Dan Duggan wrote on an incident form in June 2010 of the 'lack of gas detectors available in the control room'. Moreover, the devices were poorly calibrated, with different detectors often giving varying readings in the same location.

'Think we need a calibration standard for our miniwarns [gas detectors],' one miner reported in an incident report in the middle of 2010. Three men had measured the air in one place and come up with three different readings. 'Also there's fuck all available when ya come on shift and they're set differently,' the miner added. Another miner reported in July 2010 that he had got four different readings from four different gas detectors.[28]

On occasion even those who were carrying detectors ignored their messages of warning. The law requires miners to withdraw from a mine if flammable gas reaches two percent or more in the general body of air. Workers at Pike sometimes carried on in concentrations of two percent, and even more than five percent – within the explosive range.[29]

Gas detectors were fitted to some items of machinery, automatically shutting the motor down if methane reached 1.25 percent. Sometimes these devices were simply overridden by workers; bypassing of safety mechanisms was complained about by deputies in their official reports. 'The CH_4 (methane) valve has been broken to allow the system to be … bypassed,' one wrote. 'ASAP: Attempt to find out how and why and stop people from overriding safety circuits (PLEASE).'

That report provoked action. There was a 'toolbox talk' on the topic and efforts were made to source a better locking mechanism. But many other reports documenting the bypassing of safety mechanisms would simply disappear, or files would be closed with no action at all.[30]

In some cases the machine-mounted sensors weren't checked to make sure they were working. The in-seam drill rig operated by contractor Valley Longwall had a gas sensor that was supposed to be inspected weekly by Pike, but from July to November 2010 it wasn't checked or calibrated. The in-seam drill holes released enormous volumes of methane yet the accuracy of the sensor couldn't be relied upon.

Doug White's exposure to the information being passed up the line through the deputy's reports was as haphazard as everything else about the mine's gas and ventilation management systems. There was no systematic process whereby critical information was drawn to his attention, and he only occasionally read the deputies' daily reports where instances of gas spikes or safety breaches were recorded. On three occasions, when he did become aware of workers bypassing safety devices, he spoke to mine officials about the problem. But in a climate of white-hot pressure to produce coal, in which men had a juicy bonus dangled in front of them if they met the target for getting the hydro-mining system up and running, it was a wholly inadequate response.

There was some high-level attention paid to methane readings, however. Soon after he started working at Pike in the technical services department in April 2010, mining engineer Greg Borichevsky began getting printouts of gas levels and reporting on them at daily production meetings. If any gas spikes had occurred, he would try to find out the cause.[31]

But Borichevsky's daily reporting on methane spikes came to a stop when a new man, Steve Ellis, was hired as Pike's production manager on October 1, 2010 – just as Doug White was battling to get the new underground fan commissioned and Nishioka was trying to bring the hydro-mining system into operation. It was a time of great pressure on people and infrastructure; everyone underground was aware that Pike urgently needed to get a shipload of coal out by December.

Although Ellis had not yet obtained the necessary qualifications to fulfil the role of statutory mine manager, White effectively handed the role over to him. Ellis began running the daily meetings, and the emphasis turned even more sharply to production matters. Although methane levels in the ventilation shaft were regularly exceeding 1.5 percent, and sometimes went as high as three percent – indicating much higher, and potentially explosive, levels further into the mine – the safety of the atmosphere received scant attention at the morning meetings, according to Borichevsky. 'There was very little discussion of methane levels at the face, intersection of in-seam boreholes, and methane flows out of the main returns and fans. ... The main thrust of production meetings was on achieving target metres and tonnages and addressing any issues that were hindering production.'[32]

By the time the rush was on to get hydro mining into full production it was a year since Nick Gribble and Adrian Couchman had abandoned their trial to climb up the ventilation shaft ladder to the surface. Everyone at the mine knew the 111-metre shaft was an exhausting and demanding climb – exceptionally difficult even for a fit man, and almost certainly impossible for any worker wearing breathing apparatus while fleeing an emergency. As a stopgap measure it had been agreed that an airtight refuge would be established underground until a suitable walk-out exit could be constructed. But by November 2010, Pike had neither a fit-for-purpose refuge station where workers could safely hunker down in an emergency, nor a second exitway.

The mines inspector, Kevin Poynter, was unhappy about the ventilation shaft being used as an emergency exit, but following a visit in August 2010 he noted that it 'allows the evacuation of employees one at a time up the latter way and while this meets the minimum requirement it is agreed that a new egress should be established as soon as possible.'[33]

Workers continued to agitate for a solution, but in September 2010 the staff health and safety committee was told it would still be months before a proper second egress was established. The committee replied that this was not adequate. It wanted a firm plan of action. At a meeting a month later members were fobbed off with the news that Ellis had 'taken ownership' of the issue and would report on plans for a second egress at the November meeting.

One the first tasks Greg Borichevsky had been given when he came to Pike was to work out where a suitable second egress could be created. At the end of October 2010, he wrote Doug White a memo describing the most practical location, near the side of the aptly named Egress Creek. But driving a roadway to the spot would be neither quick nor easy: in-seam drilling had shown there were several faults along the way that could prove tricky to penetrate. He estimated the work could be done and a suitable second egress established sometime between June and September 2011 – as much as 13 months away.

At four o'clock on the morning of October 30, a section of the roof in the hydro-mining area collapsed unexpectedly. The hydro monitor was buried, a temporary stopping was blown over by the force of air, the ventilation circuit in the area was blocked, and methane rose into the explosive range.

Stephen Wylie was the deputy on duty. He and his team repaired the stopping and he wrote an incident report. It was a major event, but Wylie's report appears to have triggered no investigation into why it had happened and what the implications were for safety.

There was nothing unusual about Pike's casual response to such a serious event. By late 2010 there was a well-established pattern where incident reports would pile up on the desks of supervisors and departmental managers, to the point where the safety and training manager, Neville Rockhouse, would complain at management meetings about the lack of action. Peter Whittall would then apply pressure to have the incidents dealt with. At one stage the engineering department had 200 unresolved incident or accident reports on its books, some of them over a year old. Often, when there was a drive to 'close out' the backlog of incidents, they would be simply signed off, without any action being taken.[34]

In October 2010 it was once again time to address the logjam of unresolved incidents. Rockhouse and training manager Adrian Couchman raised the matter with Doug White. White suggested getting 'current backlog cleared and then we would start afresh with ... new management.'[35] By then, White had been appointed general manager of the mine in the wake of Whittall's elevation to chief executive, and Ellis had arrived as production manager and de facto mine manager. Whatever lessons may have been learned from proper analysis and investigation of these reports from the coalface were lost in the spring-clean that followed.

Similarly, the warnings that lurked in the many deputies' reports were seemingly ignored. From the start of October until November 19, 2010, miners reported 21 instances where methane rose to explosive levels, and 27 instances of lower, but still potentially dangerous, volumes.[36]

Through the early weeks of November the men recorded almost daily evidence of trouble. 'Cutter head tripping out on gas,' a deputy wrote on November 4. The next day another man complained of having to work in an area with exposed in-seam boreholes, which 'put me and my men at risk'. A few hours later the night shift crew recorded: 'Hit gas hole at face.'

On November 8 there were 'continuous CH_4 trips'; on November 10 there was 'Leakage of gas drainage line ... adding to problems.' On November 11 there was a report of 'Valley Longwall free-venting high levels of CH_4 through their gas drainage manifold on arrival this valve should have been turned off before those guys left their rig [sic].' On November 14: '+5 percent in area A [heading] + B [heading].'

No one was putting it all together to form a coherent picture of catastrophic danger.

On November 15, 2010, many of Pike's supportive shareholders gathered at the mine site for the annual general meeting. Some had the opportunity to go underground. The company's recently appointed chief executive Peter Whittall gave investors a detailed rundown of the many challenges faced and milestones reached over the previous 12 months: the significance of finally getting through the stone graben seven months earlier, the achievement of a second shipment of coal in September, the unreliability of the Waratah continuous miners and the critical decision to bring in the 'superb' ABM20. There was thinly veiled criticism of the departed Gordon Ward, with Whittall advising that from now on forecasts would be 'achievable' rather than 'best case' scenarios. Previously, he conceded, Pike's forecasts had 'consistently proved to be at the upper end of the range while our operations performed at the lower end.' It was now time to be 'more realistic'.

Whittall made special mention of the commissioning of the hydro-mining system, calling it a very significant milestone. 'I am very pleased with the way the process has gone. There have been no significant issues and the hydro system cuts and flows through the coal preparation plant as it is supposed to.'[37]

The shareholders didn't hear of the dreadful rates of production being achieved at the hydro panel, which George Mason had described as 'untenable' in an internal email only a fortnight earlier.[38] Nor was there any reference to the fears expressed by Nishioka, nor the recent unplanned roof fall, nor the frustrated efforts of workers like Stephen Wylie to seek proper training for himself and his crew.

And no one could have known that Doug White had, by then, decided to get out. He had been dissatisfied for months, in part because only half of the middle managers at the site reported to him, despite his key roles as mine statutory manager and operations manager. The other half reported directly to Whittall in Wellington. The arrangement wasn't functional in White's view, with important information bypassing him and flowing directly to Whittall. The arrangement changed in October when he was appointed general manager, but by then he had come close to quitting for another job in Australia.

In the event he had decided to stay put and try to make a go of things at Pike, but on November 14, the day before the annual general meeting, he emailed Australian mining headhunter Gary McCure, asking him to be on the lookout for another position. In the subject line he wrote: 'They won't be making me the scapegoat.'

'The decision to stay at Pike may well have backfired,' White wrote. 'I decided to stay because I firmly believe that the place can be successful and I was given more autonomy and control of the whole site (no increase in remuneration though). My decision was all about Pike and my family and less about me.' But in the previous two days he had seen the 'true colours of the senior leadership here and [I] don't like what I have seen.'

White was also miffed that he had received only a 2.5 percent bonus, while others he believed less deserving had received 10 percent. 'It would appear that hard work and effort, increased standards, increased productivity, increased safety performance (all of these things driven by and implemented by me) are no measure of success.'[39]

The prompt for White's email was an altercation with Whittall, who had accused him of causing a seven-cent drop in Pike's share price. Whittall had demanded to know what White had told a group of financial analysts he had escorted underground the previous week. White replied that he had given 'honest answers' to the analysts' questions, discussed possible remedies to production problems, and mentioned that 'it was the first time in 30-odd years that I'd been stumped for an answer'.[40] Indeed, in the previous few days, White had been consulting with Masaoki Nishioka by email about the chronic underperformance of the hydro system, and the day before had received a lengthy email from project manager Terry Moynihan complaining that the hydro project was being hamstrung by inadequate information about the coal seam, poor data collection and analysis, and blurred lines of responsibility among the hydro team.[41]

Whittall told White that his comments to the analysts were enough to trigger the share price fall. Chair John Dow was also annoyed. He thought White's comments were 'unguarded and relatively commercially unsophisticated', given that the company was about to go to the financial markets for yet another capital-raising.

White was shocked at the accusation. He was angered further when he consulted the share price charts and found the value of Pike stock had been falling before the meeting with the analysts. 'I am not prepared to waste my time here any longer,' White wrote to McCure.

In fact, the share price was falling for more substantial reasons than anything White had said. The previous month Whittall had announced a major downgrade in expected production, and there was a dawning awareness among some in the financial community that Pike might never achieve its long-promised production targets.

One of the analysts taken underground by White on November 11 was Andrew Harvey-Green of Forsyth Barr. The group also included a Forsyth Barr salesperson and representatives of three large institutional investors, including pension funds. They were taken to see the hydro monitor (it wasn't operating at the time), and to observe the AMB20 continuous miner excavating a roadway.

During the visit White's personal methane detector began flashing. This didn't greatly concern Harvey-Green, but White was apologetic – he had just been telling the group about the safety improvements that had been made at the mine and his determination to match Queensland's strict mine safety standards.

Whittall also happened to be at the mine site that day, and the group had a lunch meeting with him after they returned to the surface. They discussed production rates, and Whittall spoke of 'teething problems' with the hydro system.

Harvey-Green had previously rated Pike positively as an investment prospect, recommending his clients accumulate the shares. But the visit of November 11 left him with a more negative impression. 'It became clear that the issues were not going to be easy to resolve and may not be solvable at all, and that production rates may be lower and stay lower than previously thought. ... Pike had been disappointing for a long period and it was hard to see the signs of turnaround.'[42]

A few days later he published a research note for clients suggesting they 'hold' their shares, rather than buy. It signalled a polite but marked shift in perception. Pike's production rates were only a quarter of the rate required, he wrote, and it was looking unlikely that the mine would produce enough coal to fill a ship in December. That would have implications for Pike's financing requirements. Harvey-Green speculated that it might need to raise another $40 to $50 million.

In fact, Pike needed much more than that. The company was about to launch its fourth capital-raising since the IPO of 2007, and this time it was looking for $70 million.

Any hint that the operation was struggling to get its main form of coal extraction up and running, let alone that it faced myriad problems such as frequent methane spikes, a dysfunctional gas monitoring system, or the possibility of closure by the mines inspector because it still didn't have a proper second exit, would have been a blow to Pike's financial cause.

By mid November New Zealand Oil & Gas was wary of sinking more money into the project. It had a well-developed plan to get shot of its Pike shareholding; the only reason to inject more money in the meantime was to keep the company propped up until it reached production, or to avoid having its influence diluted by the introduction of other large shareholders.

But Pike and its advisers were confident they would once again find investors keen to support the mine. Finance house UBS was managing and underwriting the $70 million capital-raising; it planned to start approaching large investors in the week beginning November 22.

Despite the strained relationship between Pike and New Zealand Oil & Gas and the project's history of delay and underperformance, there was good reason to believe the money Pike needed to complete the development and get into production was out there. The world economy had survived the Global Financial Crisis, Chinese demand for coking coal was again booming, and the international spot price had reached about US$230 and was expected to go to US$300 or more over the coming few months. Investors were certain to want to grab that opportunity.

By November 18, 2010, there was just enough progress at Pike to reassure some workers that the operation could yet be a success, but not enough to ease the dark fears held by others.

The ABM20 was on the job and cutting efficiently. The new fan was running more smoothly and generating a welcome increase in the volume of air flowing through the working places of the mine – indeed, it created a breeze in the hydro panel 'cold enough to freeze your snot', as one man put it. Doug White was still respected as a manager who cared about the men, and was seen to have the competence to turn Pike into a success.

But Queensland miner Willie Joynson was among those who wanted out. He had been at Pike for 15 months. Nicknamed Digger, he was 49 and had first worked in a mine as a 17 year old. He was sought after by mining companies because of his long experience and capacity for hard work. He had shifted with his wife Kim and their children to the West Coast because it offered a better life than northern Queensland – he could go to work underground and be home for tea each night, rather than having to be away at distant mines for days on end. He also liked the idea of working in a small mine.

When he first got to Pike he had some reservations about the place: he thought the company's expectations were too high for the size of the mine, and there was a need for better training and more experienced workers. However, he thought it had the potential to grow and develop.

But from July 2010 Joynson started expressing concerns to Kim and to friends. He thought Pike was a time bomb just waiting to happen. He talked about problems with gas, about the fan that kept tripping off, machinery that wasn't fixed correctly or quickly, and the frustrations caused by previous crews who didn't set things up properly for the follow-on crew (mainly because of machinery breaking down). He spoke of the number of people who were leaving, and others who just didn't turn up for their shifts.

He and Kim decided he should leave, and he found another job in Australia in August. Then his fears were calmed after talking to Pike managers. The couple decided to stay a bit longer and give things another go.

In September Joynson again started expressing anxiety about gas, machinery and ventilation. He worried about the number of contractors underground who, while good at their trades, had no experience of working in coal mines. 'He would say, "There are so many more factors to deal with working underground than working in a house, and that safety should not be taken for granted. You cannot get experience on paper, you can only learn from doing the job in the correct conditions and from listening to those who have been doing it a long time,"' Kim would later recall.

The tradesmen were keen to learn and wanted more on-the-job training, but Joynson felt managers weren't listening to their concerns.

In late October he told Kim things were not very good. 'In particular he said that if something happened he would not be coming home. I said he should go to Australia, whatever the response from future employers might be to him just walking out.'

The couple decided they would move back to Australia in time for the children to start new schools in February 2011, the beginning of the academic year.

In November he told Kim there had been a near miss at work. She had never seen him look so worried. 'When Willie is stressed he sleepwalks. In our marriage this has only occurred about five times, twice on family issues and three times in New Zealand. In July 2010 he was doing it. This is why I booked his flight home. I felt he needed to get away – I felt Pike was the cause of his stress.'

Then in November he sleepwalked again. 'One time I woke up and he was in the wardrobe talking about the mine. His concerns were so strong he told me that if anything happened they would not get out alive. He felt there was no clear way out: the emergency exit was no good, some could not fit up the ladder, and it was too far by foot.'

On the weekend of November 13 he spoke of another near miss. Kim again told him to quit and walk away. 'Willie said he could not leave the guys at that point, that I did not understand, and that the inspectors were coming to look into things.'

Over the following week there was not much time to talk as both Kim and Willie were working shifts. When they did find time, Kim would become angry and tell him to 'just leave Pike', which didn't ease his stress.

'Maybe if I had taken more notice, the signs that something was really wrong were all there.'[43]

They were signs that others, who had the power to act, failed to see.

NINE
Who Will Say Stop?

It was almost two decades since Harry Bell, then chief inspector of coal mines, had brought his regulatory authority to bear at the Huntly West underground mine. In September 1992 he had ordered the place shut, thus saving the lives of dozens of workers when it exploded three days later.

In the intervening period, the skilled and robust inspectorate that Bell had served for 15 years and led for two had been all but destroyed. By November 2010, underground coal mines such as Pike River were effectively left to run their operations however they chose, provided they were seen to work within the vague and elastic framework of rules that had emerged from a tsunami of deregulation in the early 1990s.

As a young regional inspector overseeing the underground mines of the West Coast in the 1970s and 1980s, Bell had been under an obligation to visit large gassy underground coal mines weekly and smaller mines monthly. His job was to ensure compliance with the 1979 Coal Mines Act, a prescriptive and detailed rulebook written in the blood of miners killed at Strongman, Dobson, Brunner and other miners' graveyards over a century of coalmining in New Zealand.

At every inspection Bell would first look at the gas book, in which mine officials were obliged to record every instance of methane detected. (The gas book was also the first point of daily reference for a mine manager.) Bell would then walk through the mine and talk to the workers, the workmen's safety inspector and the mine manager. He'd take his own gas readings, inspect stoppings, and assess ventilation. If a mine was struggling with a problem, he would suggest solutions based on his decades of experience as a miner, manager and regulator.

If an application to develop a new coal mine in his district was lodged, the document would be sent to him for detailed review. As well as checking the technical viability of the application – the proposed ventilation system, the method of extraction, and so on – he would look into the developer's financial credibility. It was well understood that if a company attempted to mine on the basis of inadequate exploration or flawed methodology, or ran short of money and took safety shortcuts to win coal, there was a high chance things would go tragically wrong.[1]

But by the time New Zealand Oil & Gas applied for a licence to develop an underground coal mine at Pike River in 1996, the mining inspectorate was in the process of being dismembered. It had no say at all in whether NZOG's scanty 28-page application ought to be granted. That role had become the sole preserve of a unit of government called Crown Minerals. And Crown Minerals had no interest in whether Pike's proposal was technically sound, financially viable or safe; it called for no information on its capital requirements or expected profitability.[2]

Had NZOG's application for a permit to mine at Pike River come across Bell's desk when he was an inspector, he'd have recommended it be declined on the basis that there wasn't enough information to properly assess whether it could be developed according to good mining practice. As it happened, Crown Minerals served simply as a permitting factory. In 1997 it gave NZOG the licence.

By the time Pike River Coal Ltd was tunnelling towards the coal seam ten years later, it had crafted a highly polished corporate profile as a modern resources company that would extract high-value coal from the pristine wilderness with surgical precision and safety. It promised to be an industry leader that would set new standards of excellence in an industry hidebound by tradition.

The inspector responsible for checking on Pike's health and safety during the early phase of the project was Michael Firmin. At the time, Firmin was the sole mining inspector in New Zealand, responsible not only for underground coal mines but also quarries, tunnels and opencast mines – about a thousand workplaces in all. His superiors in the Department of Labour had no knowledge of underground coal mining, and the post of chief inspector of coal mines, once filled by Harry Bell, had been abolished.

Gone, too, was the obligation to inspect underground coal mines frequently. Firmin's task was to make three-monthly visits, which were generally prearranged with the mining company. If he wished to inspect a mine more often than that, his bosses were inclined to ask: 'Why do you need to go to these places this often?'[3]

Firmin made his first inspection of the Pike site in 2007, when the McConnell Dowell crew were still slowly boring the tunnel through crumbly rock. He was shown a PowerPoint presentation of the project by Peter Whittall, who mentioned that the company intended to put its main ventilation fan underground. By then, the detailed coal mining legislation that had guided the inspectors of Bell's generation had been thrown out, replaced with the Health and Safety in Employment Act and its supporting Mining Underground Regulations 1999.

Firmin consulted the regulations and concluded there was nothing to prevent Pike putting the main fan underground. Whittall, too, assured him there was 'nothing that would stop Pike River doing this'.[4]

Pike's public relations efforts over the previous few years had won over not only investors, financial analysts and politicians. They had also helped convince Firmin that Pike was a company with well-developed health and safety practices. 'They were willing to comply, wanting to perform, wanting to be involved with best practice.'

Pike was always responsive to his queries, and seemed committed to safety.[5] It was thus the sort of employer to whom the Department of Labour was reluctant to apply a heavy regulatory hand. Rather than use its legal powers of enforcement to stop unsafe work, or prosecute for poor health and safety practices, the department preferred to negotiate solutions.

Firmin was reassured about the proposal to locate the main fan underground because Whittall told him there would be a back-up fan on the surface that would kick in whenever the main fan stopped. And the underground fan would be positioned in such a way as to protect it from the path of an explosion, should such an unlikely event occur. He therefore mounted no challenge to Pike's unconventional plan.

In April 2008 the size of the mining inspectorate doubled: Firmin was joined by a trainee inspector, Kevin Poynter, who took over responsibility for policing safety at Pike. Poynter was an experienced mine manager, but in line with the department's policy he, too, took a light-handed approach with Pike. It continued to be seen as a 'reasonably compliant employer' and a company striving for 'best practice'.[6]

Poynter was not dissuaded from this view despite a succession of incidents and accidents at the mine during the 28 months he monitored the site. In November 2008 there were the ten gas ignitions that had terrified the workers. Under the regulations, Pike was obliged to notify the inspector of each one, but it let Poynter know of only four of the incidents. He found out about the rest only because Harry Bell blew the whistle.

Then there was the collapse of the ventilation shaft in early 2009. This might have prompted a more sceptical view of Pike's competence to manage the geological risks of its ambitious project, but appears not to have damaged Pike's brand as an exemplary outfit. And in early 2010 Poynter was called to investigate a serious accident at the mine in which a worker's foot was severely crushed and partially degloved, the skin and subcutaneous tissue torn off. In the course of the investigation it was found that a safety mechanism on a machine had been deliberately bypassed.[7]

None of these incidents prompted Poynter to revise his view of Pike. His attitude may perhaps have been different had be not been so overburdened by the scale of his job and so poorly supported by his employer, the Department of Labour. Without fail, Poynter issued a monthly cry for help on his report to his superiors, writing under 'Issues and Risks': 'With only two warranted inspectors covering the country, resources are extremely stretched. In addition there is a lack of knowledge or inspections of high-risk extraction sites throughout the lower half of the North Island.'[8] He made repeated pleas for a third inspector, for the reinstatement of a chief inspector of mines who understood the industry, and for proper training.

At one meeting of the Mining Steering Group – a ragbag group of officials formed to lend support to the lonely mining inspectors – Poynter noted that the chief inspector of mines in Tasmania had told his masters just before the 2006 Beaconsfield gold mine disaster, in which one man was killed and two trapped underground for two weeks, that he was 'not in a position to provide an adequate inspection service with the resources at his disposal'. Poynter's implication was plain: an inadequately resourced inspectorate could open the way for a similar tragedy in New Zealand.[9]

The steering group attempted to get the message across to those higher up the departmental hierarchy. 'Should there be deaths or catastrophic failure, as can occur in mining, questions of our current staffing and regime will be asked. We need to be comfortable with the investment decisions made,' the group advised in a February 2010 request for a third inspector.[10] The request was rejected, although departmental officials made a file note that the organisation would look bad if the worst happened.

In March 2010 the following item was lodged in the department's risk register: 'Limited mining resource. May have service failure, certainly very constrained service. Reputational risk in an event.'[11]

Despite his formidable workload, Poynter was saddled with additional tasks, including accompanying the minister of labour, Kate Wilkinson, on an underground visit to Pike in mid 2009. A few months later the company's annual review sported a photo of a smiling Wilkinson kitted out in protective clothing in readiness for a trip underground with Peter Whittall and other members of Pike's management team, an image that helped maintain the company's brand as a standard-setting modern miner.

On November 2, 2010 Poynter made his fourth inspection of Pike River mine that year. No one told him, and he didn't ask, about the recurring problem of methane spikes virtually every time the hydro monitor started cutting coal. He didn't see, and wasn't shown, the readings from the one functioning methane sensor measuring the contaminated air leaving the mine, which indicated that gas levels had risen frequently into the explosive range over previous weeks. Pike was obliged under the underground mining regulations to report to him every instance of uncontrolled methane accumulations. It had not done so.

Poynter similarly had no idea that the mine's gas sensors were not being calibrated regularly, nor that electrical machinery was often tripping out because of high methane levels. He wasn't told of incidents where cigarette butts, lighters and aluminium drink cans – all prohibited items that could provide a spark sufficient to ignite explosive methane – had been found underground. Nor was he told of further instances where safety devices on machinery had been overridden by workers. He didn't read the incident book, which was littered with workers' reports on safety issues.

The information was all there, buried in the organisational dead ends of a failing organisation, but it wasn't readily apparent to a time-poor and underpowered inspector. And where problems such as the mine's continued failure to construct a proper emergency exit were obvious, Pike's profile as a cooperative and well-intentioned employer guaranteed a muted regulatory response.

Poynter was well aware that the ladder up the ventilation shaft was the only way workers could get out of the mine if the main tunnel were blocked by fire or rock fall, and he knew it was unsatisfactory. In April 2010 he had asked the company to provide a plan for the development of a proper walk-out exit. Pike management assured him the matter was a priority, but it was six months before the plan was produced. Drawn up by Greg Borichevsky, it pinpointed the location of the proposed exit, outlined the roadway development needed to reach that part of the mine, and estimated it could take a year or more to complete it.

Meantime Pike was pulling out all stops to ramp up the hydro-mining system. Getting coal out was more important than building a usable emergency exit for the workers.

Poynter contemplated using his powers of enforcement over the issue, but thought such a move would be difficult. The law was vague. The regulations stated that mine employers had to 'take all practicable steps to ensure that every mine ... has suitable and sufficient outlets, providing means of entry and exit for every employee in the mine.' But what on earth did that mean? He thought that even though it was unsuitable the ladder probably met 'minimum requirements' – but told Pike a new egress needed to be established as soon as possible.[12]

Even if he did issue a prohibition notice against Pike, forcing mining to stop until the second exit was constructed, Poynter thought the company would challenge the Department of Labour in court. After all, it was humanly possible to climb the ventilation shaft and he had been assured that people had done it. In any event, Poynter thought it most unlikely that his superiors would support his flexing his legal powers to shut down the mine. 'With the profile that Pike River had, I did have concern that I wouldn't have got the support.'[13]

By November 19, 2010, neither Poynter nor Firmin had raised a single regulatory stick against Pike, and the mine's march towards catastrophe was uninterrupted.

Pike's six-man board of directors held its monthly meeting at the mine site on November 15, 2010, just before shareholders were due to assemble for the annual general meeting. Only three of the six directors were physically present. Tony Radford dialled in by phone from Sydney, and Arun Jagatramka and Dipak Agarwalla sent their apologies. There was nothing unusual about the poor turn-out. Radford usually attended by teleconference and seldom made it to meetings at the mine site.

However, the level of board participation by the Indian directors was of particular concern to the Pike chair, John Dow. The board had gone to some lengths to accommodate the geographical spread of directors:

meetings were an equal mix of in-person and telephone conferencing. In 2010, meetings had so far been held in Sydney, in Wellington and at Gujarat's mine site in Wollongong, New South Wales. Dow had also encouraged the Indians to appoint alternate representatives, which they did. But by October 2010 he still harboured concerns.[14]

The first item of business was, as usual, finance. The company was about to start its fourth capital-raising since the IPO, this time for $70 million, and had reputable finance house UBS on its side.

Dow thought it was also 'timely' for the board to focus on health, safety and environmental matters at the meeting, and asked Doug White in for a briefing. It was the first time the board had ever called on White to answer questions about topics such as gas, ventilation and risk management. He delivered a reassuring overview. The mine, White said, was following the stringent Queensland recommendations for dealing with gas emissions, and the new underground ventilation fan – which had been running smoothly for only a few days – had greatly increased the amount of air moving through the mine. Methane in the gassier part of the mine was being drained away, and with 'adequate ventilation' it was 'more a nuisance' than a barrier to operations.

White advised that procedures existed for evacuating workers in the event of ventilation failure, and said a second fan and emergency exit were to be installed the following year. In the meantime the ventilation shaft 'provided a means of emergency egress' and there were plans to increase the size of the underground refuge station.[15] Nothing in White's reported comments revealed the extent of his disenchantment with Whittall, whom he referred to the next day, in an email to a friend in Australia, as a 'dodgy git'. He found it very hard, he wrote to his friend, 'to work for someone who has made or overseen so many stuff ups and blames everyone else.'[16]

The board didn't ask for, and wasn't given, the information that would have flashed warning lights about the mine's safety management systems failing. Members took White at his word that methane levels and ventilation were under control; they had no system for independently verifying that the mine's most significant danger was being properly managed. Aside from White's verbal assurances, their primary source of information on safety at the mine was statistics that set out the number of accidents that had required workers to seek medical treatment – sprained ankles, chipped teeth, cut fingers and the like. They had no way of knowing how often the atmosphere in the mine was in a potentially explosive state.

The board had a subcommittee dedicated to health, safety and environmental issues. It was chaired by Dow. It hadn't met for 13 months. There were plans for third-party audits to help the board assure itself that Pike's safety and risk management systems were up to scratch. This was an excellent idea, particularly as, with the exception of Jagatramka, none of the directors had any experience in the underground coal mining industry. Jagatramka's company Gujarat NRE had two underground coal mines in New South Wales, and Jagatramka had been touted as a visionary of rare distinction who had led the growth of Gujarat NRE into Australia. Unfortunately, he was frequently absent from Pike's board meetings.

By November 2010 the board had still not undertaken any third-party audits. Dow was in favour of such measures, but management had argued that outside audits would be more useful after the company's entire safety management system was fully implemented and in operation.[17] In the meantime, the board appeared to have reasonable grounds for assuming the safety of the workforce was uppermost in the minds of the company's management. Peter Whittall had just a few months earlier taken a leading role in the establishment of a high profile organisation aimed at improving workplace health and safety throughout New Zealand. The Business Leaders Health and Safety Forum, of which he was a steering group member, had been launched by Prime Minister John Key in July 2010, and celebrated as the first of its kind in the world.[18]

There was also the reassuring presence of Neville Rockhouse on Pike's management team. Rockhouse had been the mine's safety and training manager since 2006. He was a qualified health and safety professional with many years' experience. And while he had never worked at a gassy mine such as Pike before, nor developed a health and safety system for an underground mine from scratch, he shared the dream of making Pike the best new mine in the country. Two of his three sons worked underground in the mine.

The board failed to detect that Rockhouse was struggling to implement the safety rules and procedures he was tasked with designing, and was so underresourced he barely had time to go underground to check out things for himself. A colleague, Michelle Gillman, feared he would have a heart attack or breakdown because of the stress he was under.[19]

Rockhouse asked often for more staff and resources to help develop the mine's health and safety systems, mostly without success. His relationship with Whittall was tense. On one occasion the Pike boss humiliated him in front of his peers while he was giving a presentation on hazard identification. Whittall had objected to a couple of typographical mistakes, and when Rockhouse couldn't see them Whittall rose to his feet and began yelling and slapping at the wall to indicate the location of the errors on the screen. Rockhouse was so shocked he tried to resign – not for the first time – but was dissuaded by his management colleagues.[20]

By October 2010 Neville Rockhouse had overseen the creation of 386 documents setting out a wide range of safety-related procedures – among them how to operate machines safely, standard responses to serious incidents, risk assessments and management plans. It was up to the manager of each specialist department – engineering, technical services, environmental and production – to sign off documents related to their sphere of influence. Less than a third had been signed and finalised. Rockhouse tried to help managers develop their safety documentation but Whittall rebuked him, saying on one occasion, 'Keep your bloody nose out of it.'[21]

The board was unlikely to have known of such strains: the operational information it received from the mine was channelled through Whittall. Indeed, Rockhouse had been told by Whittall not to send emails directly to board chair John Dow.[22]

Nevertheless, it was the task of the directors to set the company's direction and ensure the risks were being properly managed by the executives they appointed. And while they might struggle from the outside to discern interpersonal friction, such as that between Whittall and Rockhouse, there were other portents flashing under their collective noses. The enormous turnover of senior managers – including men such as Kobus Louw, who left after a short time despite having gone through the stressful business of relocating his young family from the other side of the world – pointed to a deep malaise. In four years the project had burned through four technical services managers – the fourth, Pieter van Rooyen, quit early in November 2010 – and was on to its third engineering manager. And in the two short years since the project had reached the coal seam, the critical position of statutory mine manager had changed hands six times.

None of this appeared to provoke alarm around the board table. It was simply assumed that the rapid churn was a result of the Australian mining boom and the tough competition for skilled mining personnel.[23]

Nor did alarm bells ring about the extraordinary step taken by Pike's former project manager, Les McCracken, when he warned John Dow in August 2009 that the mine had serious problems with morale and leadership. Dow had told Whittall and Gordon Ward about his conversation with McCracken, and suggested they bring in experienced mining consultant Dave Stewart to help out, but he then had no further involvement in the matter. He considered the issues raised by McCracken to be 'routine' and not something the board needed to get involved in. So he didn't ask for, and wasn't given, the detailed reports from the coalface that Stewart completed at Pike in early 2010.

Had the directors read Stewart's work they may have been dissuaded from the assumption that all was well at the mine. Stewart's reports documented inadequacies with Pike's gas detection and ventilation systems, pointed out that the ventilation shaft was not a suitable emergency exit, and warned that workers felt their reports on safety incidents were being ignored by management.

Nor did the board ask for the results of a comprehensive risk survey done in mid 2010 on behalf of Pike's insurer by Hawcroft Consulting.

Dow knew the work was being done but thought that this, too, was the preserve of management. Had the board been sufficiently curious to read the report it might have been moved to undertake more rigorous surveillance of risk management at the mine.

Hawcroft assessed the management of methane at Pike to be of a low standard and said general housekeeping was poor, with 'equipment, parts and rubbish evident'. It noted there were unknown risks associated with hydro mining – in particular the chance of a roof cave-in. And it judged Pike's general risk management to be below average.[24]

Pike's directors didn't receive the Hawcroft report. Dow considered it dealt with matters that were 'operational' and therefore the domain of the company's managers. He thought directors needed to be careful not to dabble in the operational side of the business; the board's job was governance, not management. He likened the relationship between the board and managers to that of 'church and state'.[25]

And so by November 2010 Pike River Coal's board of directors continued to believe, despite so many signs to the contrary, that everything was under control at the mine.

Pike River mine was awash with information foretelling catastrophe, but all those who had the power to act on the warning signs were deaf and blind to them. Vital information lay fallow on desks and in files, and pleas from men at the coalface for action and improvements went unheard and unanswered. Even Doug White, Queensland's former deputy chief inspector of coal mines and a man admired by his staff for his efforts to improve safety at Pike, couldn't see the dangers that were keeping Willie Joynson awake at night.

After Peter Whittall moved to Wellington in early 2010, White was the most senior manager at the mine site. As operations manager he had overall responsibility for production, engineering, health and safety, and coal processing. He also absorbed the position of statutory mine manager in June 2010, a role that carried the legal duty to personally supervise the health and safety aspects of the operation.

Yet White had an incomplete picture of what was really going on. He didn't know the fixed sensors in the mine's return roadways that were supposed to transmit accurate and continuous methane readings to the control room were either not working, had been disconnected from the control room, or were inaccurate. He assumed the sensors were being regularly calibrated and maintained, but they were not.[26]

After Steve Ellis arrived at Pike to take up the job of production manager at the start of October 2010, White effectively handed over his duties as mine manager to the new man, even though Ellis did not have the right qualifications to hold the job. Until that time White had been seeing regular updates on methane trends in the mine because Greg Borichevsky, the technical services coordinator, brought the information to daily production meetings. White assumed that this flow of information carried on when Ellis took command of those meetings, but it did not.

White was conscious of the need for better training at Pike, and engaged Harry Bell to provide refresher courses on self-rescue techniques. The first course was held on a Friday afternoon in early October 2010, but the following week only three men could be spared from production to attend. After that the sessions were put on hold.

White also knew that the lack of a proper emergency exit from the mine was a matter of ongoing anxiety for the men underground, and that the issue came up repeatedly at meetings of the company's health and safety committee. The committee – a poorly attended group where there were sometimes more managers than workers in attendance – had escalated its concern about the lack of a proper second egress in September 2010, writing a formal letter to Pike management. Yet White told the board on November 15 that the ventilation shaft provided an escapeway until a second egress could be built the following year.

White was justly proud of the improvements he had made at Pike – in particular bringing in the ABM20, which had begun to turn around the mine's chronic inability to produce coal. He considered the establishment of a second exit to be a priority. But despite his determination that Pike should be run according to the stringent mining standards applicable in

Queensland – where two exits in fresh air are mandatory – he oversaw the push into hydro mining instead of suspending operations until a second escape route for the workers had been built. He knew that a mine of Pike's design 'would not have existed in Queensland'.[27]

By November 19, 2010 the absence of an emergency exit from the mine was, on its own, sufficient cause to shut down production and address the safety deficit. Doug White did not take that step.

The union representing mine workers, Engineering, Printing and Manufacturing Union (EPMU), could have brought the mine to a halt, at least temporarily, by encouraging strikes, pickets or bans over safety. But it caused no disruption to Pike's path to calamity. It had a limp presence at the mine, in part because it wasn't welcome.

There was only ever one walk-out over safety, when mine deputy Dan Herk threw down the gauntlet about the lack of mine vehicles available to quickly evacuate workers in the event of an emergency. Herk called the local EPMU representative, Matt Winter, and said he was concerned for the men's safety; Winter advised he should, therefore, walk out. Herk led the men out of the mine. Shortly afterwards Winter received an angry call from Pike's human resources manager Dick Knapp, advising him to tell the men to go back to work; when Winter refused, Knapp threatened to sue the union. The issue the men were protesting about was attended to within a matter of hours, with the prompt repair of a broken-down vehicle that had been out of action for three weeks.[28]

Winter was aware of workers' concerns about the lack of a proper emergency exit, and he had heard about the series of methane ignitions in late 2008. He was also worried about the high number of cleanskins – workers new to mining – at Pike. He understood that it was desirable in underground coal mining to have a ratio of experienced to inexperienced workers of about four to one. Pike had a much larger proportion of inexperienced men than other sites he looked after.[29]

It wasn't easy to enlist Pike workers into the union. Some told Winter they didn't want to upset management by signing up. And he got the impression Pike management wasn't interested in forming any sort of relationship with EPMU. Pike had an internal health and safety committee but the union had no representation on it. Winter found Pike management 'arrogant and unwilling to listen. They were prepared to tolerate the presence of the union in line with their legislative obligations, but they were not at all interested in developing a good relationship'.[30] He left his job in early 2010 and handed over to a new man, Garth Elliot.

Others at the site also had the impression that the company preferred not to have a strong union presence. In 2009, when health and safety manager Neville Rockhouse sought to have the union involved in a training exercise, Peter Whittall told him in an email: 'Please do not use the union in the same sentence as anything at Pike. Our relationship and the way we communicate is between us and our employees.'[31]

And so men like Willie Joynson, who went underground every day to earn a living, and who were entitled to the protection of robust safety systems and equipment that left a fat margin for error, were working on the edge. Pike River mine, which needed to have the best of everything to succeed in its tough environment – the best geological knowledge, the best equipment, the most rigorous safety regime – had the worst of everything. Joynson and his workmates were exposed on all sides by those whose job it was to protect them: a regulator that was submissive and unwilling to use the powers at its disposal; a board that was incurious, bereft of knowledge and experience of underground coal mining, and unable to see the symptoms of failure; management that was unstable, ill-equipped for the environment, and incapable of pulling together all the pieces of its own frightening picture; and a union that was marginalised and irrelevant.

The top of the ventilation shaft, showing severe damage to the surface fan infrastructure from the November 19 explosion. *Mines Rescue Service*

ABOVE Family members with Superintendent Gary Knowles, the incident controller, waiting for news of their men the day after the first explosion on November 19. *Martin Hunter*

BELOW Gary Knowles speaks to media on Sunday, November 21, flanked by Trevor Watts, Mines Rescue Service general manager, and Peter Whittall. Whittall said compressed air was being pumped into the mine and that there could be 'heating' in the coal. Next day it was established that the line had in fact ruptured 1.6 to two kilometres into the tunnel. *Fairfax Media/The Press*

ABOVE Energy and Resources Minister Gerry Brownlee and Prime Minister John Key at a press conference at Greymouth Police Station on November 22, as hope dwindles. On Key's first visit on November 20, Peter Whittall had told him some of the men could still be alive. *David White/The New Zealand Herald*

BELOW Daniel Rockhouse, one of the two men who survived the blast, hugs family members after a November 22 briefing by police and Pike River Coal officials. *Mark Mitchell/The New Zealand Herald*

ABOVE Peter Whittall, with his wife Leanne and Pike chair John Dow, addresses the media following the second explosion on November 24. *Martin Hunter*

RIGHT TOP Tag board at the Pike River mine administration office, bearing the identification tags of the 29 missing men. *Mines Rescue Service*

RIGHT The mine portal after the second explosion on November 24, which extinguished all hope for men in the workings. Efforts to seal the portal had not yet begun; the conveyor structure shown in the photo made the task very difficult. *Mines Rescue Service*

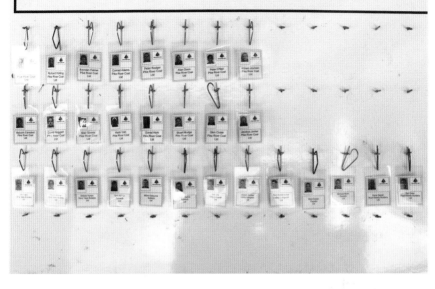

Flowers and photographs were laid near the mine portal in the days after the deadly explosions. *Michael Buckley*

The twisted remains of a stainless steel switchboard door; it had been blown from inside the mine workings 111 metres up the Alimak raise and ventilation shaft on to the forested tops of the Paparoas. Mines Rescue workers who found it not long before the fourth explosion suggested to police that the area be searched for human remains, but this did not occur until several weeks later. *MPR Collection*

ABOVE The top of the slimline shaft after the fourth explosion, November 28. Four days earlier, a laser scanning crew had been waiting here to be picked up by helicopter when a huge plume of smoke, soot, coal dust and other debris had erupted up the slimline and ventilation shaft: the mine had exploded a second time. *MPR Collection*

RIGHT TOP Smoke pours from the mine after the fourth explosion on November 28. *The New Zealand Herald*

RIGHT Flames leap from the main ventilation shaft deep in the forested Paparoa Range, November 28. The flames continued for another week as the country watched in horror. *Iain McGregor/The New Zealand Herald*

ABOVE Families (seated) at the memorial service for the 29 missing men, held at Greymouth's Omoto Racecourse on December 2, two weeks after the first explosion at the mine. Tables for each of the lost mine workers were adorned with flowers, photographs and favourite possessions. *Martin Hunter*

LEFT TOP Peter Whittall addresses the memorial service on December 2. Seated to his right are the Greymouth mayor Tony Kokshoorn, the prime minister, John Key, his wife Bronagh, and Tony Kokshoorn's wife Lynne. *Stewart Nimmo Photography*

LEFT Flames continuing to leap from the ventilation shaft a week after the fourth explosion. *Mines Rescue Service*

ABOVE Mines Rescue Service first formally proposed on November 21 that the mine be sealed, in the hope this would reduce the chance of a second explosion. Not all the assembled experts agreed. Two shipping containers were finally placed in the portal and shotcrete applied remotely two weeks later. It was very difficult to work in close proximity to the portal due to the risk of further explosions. *Mines Rescue Service*

BELOW Mines Rescue workers seal the ventilation shaft, December 12, 2010. *Mines Rescue Service*

ABOVE A memorial erected by the families of the 29 men killed in the Pike River mine disaster. Land for the memorial, which stands near the turnoff to the mine's access road, was donated by the Berry family of Atarau. Paul Berry extracted the large boulder and the 29 accompanying stones for the memorial from near the mine portal. *Stewart Nimmo Photography*

BELOW A tribute to the two survivors of the tragedy, Daniel Rockhouse and Russell Smith. In a gruelling feat of endurance, Rockhouse pulled Smith, who was semi-conscious, down the tunnel to safety. *Stewart Nimmo Photography*

ABOVE LEFT The Royal Commission on the Pike River Coal Mine Tragedy was set up by the government ten days after the first explosion and reported two years later. One who gave evidence was Harry Bell, former chief inspector of coal mines. From the beginning, he had had deep concerns about the mine's design. *Fairfax Media/The Press*

ABOVE RIGHT Robin Hughes, also a former chief inspector of coal mines, gave evidence in July 2011. An experienced mine ventilation engineer, he was the first to analyse gas samples taken after the November 19 explosion; they convinced him a methane fire was burning underground. *Fairfax Media/The Press*

LEFT TOP Dan Duggan, controller at the mine's surface control room, phoned underground just after 3.44 p.m. He spoke to Scottish engineer Malcolm Campbell briefly but there was no further response. It was another 41 minutes before mine manager Doug White gave him the go-ahead to call Mines Rescue Service. Here Duggan, whose brother Chris died in the mine, gives evidence to the Royal Commission. *Fairfax Media/The Press*

LEFT Justice Graham Panckhurst (centre) chaired the inquiry, assisted by commissioners Stewart Bell (left) and David Henry (right). The commission concluded that numerous warnings of catastrophe went unheeded at Pike, and that the mine's executive managers had a culture of production over safety. *Fairfax Media/The Press*

Bernie Monk, the spokesperson for some of the Pike River families. Monk's son Michael was killed in the mine. *Martin Hunter*

ABOVE Grey District Mayor Tony Kokshoorn at the gate to Pike River mine in March 2012. The closure of the mine dealt a body blow to the struggling West Coast community. *Martin Hunter*

BELOW Anna Osborne, whose husband Milton died in the disaster, erects a sign on the Cobden Bridge with Milton's close friend Tom Daly. After the mine was sold in July 2012, responsibility for bringing out the men's bodies shifted to the new owner, Solid Energy Ltd. *Greymouth Star*

Conrad Adams
43, Greymouth

Malcolm Campbell
25, St. Andrews, Scotland

Glenn Cruse
35, Greymouth

Allan Dixon
59, Rūnanga

Koos Jonker
47, Limpopo, South Africa

William (Willie) Joynson
49, Maryborough, QLD, Australia

Stuart (Stu) Mudge
31, Rūnanga

Peter O'Neill
55, Rūnanga

Terry Kitchin
41, Rūnanga

Samuel (Sam) Mackie
26, Christchurch

Milton (Milt) Osborne
54, Ngāhere

Joseph Dunbar
17, Christchurch

Riki (Rik) Keane
28, Greymouth

Michael Monk
23, Greymouth

John Hale
45, Hokitika

Andrew (Huck) Hurren
32, Hokitika

Christopher (Chris) Duggan
31, Dunollie

Daniel (Dan) Herk
36, Rūnanga

David (Dave) Hoggart
33, Greymouth

Richard (Rolls) Holling
41, Blackball

Brendon Palmer
27, Greymouth

Peter (Pete) Rodger
40, Perth, Scotland

Blair Sims
28, Greymouth

Keith Valli
62, Nightcaps

Benjamin (Ben) Rockhouse
21, Singleton, NSW, Australia

Joshua (Josh) Ufer
25, Charters Towers, QLD, Australia

Zen Drew (Verhoeven)
21, Greymouth

Kane Nieper
33, Greymouth

Francis Marden
41, Barrytown

Pike River Coal

Sub Tech Contracting

Valley Longwall Drilling

Boyd Kilkelly Builders

Pizzato Contracting

Chris Yeates Builders

The Pike 29, a montage created after the event. *Stewart Nimmo Photography*

ABOVE The sealed entrance to Pike River Mine as it was in mid 2013. Planning was underway to re-enter and search the stone tunnel. *Solid Energy Ltd*

LEFT Judge Jane Farish convicted Pike River Coal Ltd (in receivership) in Greymouth District Court on April 18, 2013 of nine breaches of the Health and Safety in Employment Act. She described the Pike disaster as the 'health and safety event of this generation … a worse case is hard to imagine.' *Fairfax Media/The Press*

TEN
November 19

Allan Dixon left his mother Nan's house in Rūnanga before six in the morning. He had to be down the road at Coal Creek corner to meet the Pike workers' bus, which would take almost an hour to drive up the Grey Valley and into the mountains to the mine site. As usual, he had a most excellent lunch to look forward to. Nan always prepared his crib box the night before: fresh sandwiches, sometimes a little of the salmon that he was partial to, a small tub of yoghurt, and – without fail – a range of cakes and biscuits baked in her tiny kitchen. Often she'd make an extra fruit cake or a batch of shortbread he could share with his mates; she always felt for the men who commuted from afar to work at Pike and were away from their families.

Dixon, Peter O'Neill, and Keith Valli formed one of the hydro-mining crews: they worked the 12-hour day shift starting at seven. O'Neill was also from Rūnanga, and the men's mothers were close friends. At 62, Valli was one of the oldest workers at Pike; he was from Southland and commuted home on his days off. The three were among the most experienced miners at Pike.

They took over from the night shift crew led by Stephen Wylie, which had made frustratingly slow progress using the hydro monitor to extract a bench of extremely hard coal. Wylie reported nothing of particular concern when he spoke to the incoming crew. He took methane readings and reported them to be very low – under 0.5 percent. He also checked for carbon monoxide and found no sign of the gas, which is an indicator of spontaneous combustion in the coal seam.

Another deputy, Simon Donaldson, had detected carbon monoxide eleven days earlier, recording on his statutory report: 'Noticed high levels of CO in bleeder road when ventilation was cut off. Still no CO once ventilation was restored.' Donaldson's report noted that he had spoken to George Mason about organising regular bag samples to be taken from the monitor panel. It is not clear what the outcome was, but Donaldson's report was signed off by 'SE' – probably Steve Ellis, the production manager and de facto mine manager – some time on November 19.[1]

Just after eight, the hydro team called through to the surface control room on the underground communication system to say they were cranking up the monitor. Dixon, O'Neill and Valli continued cutting coal until about 11.30, when a strainer on a pump became blocked. Then at 12.20 there was a planned shut-down in water to the mine so some repairs could be done at the coal preparation plant; no further mining with the hydro monitor would be possible until the water was turned back on.

Milton Osborne left his home in rural Ngāhere at 5.30 that morning. His wife Anna and their two teenage children Alisha and Robin were still asleep; without waking her, he gave Anna his usual kiss on the cheek and slipped out the door. He hadn't got home from work the night before until about nine-thirty or ten, so they had barely had time to catch up. The children hadn't seen much of him at all over previous two weeks because of the huge hours he was working. He was contracted to install fluming pipes underground, and there was a big push to get everything in place so the mine could get into steady production. He was giving the project his all, but Anna worried about the cost to his health – he had been uncharacteristically grumpy over recent weeks, and his skin had taken on such a pallor she had told him, 'You look sick, Milt.'

Anna worried about the gases underground, but was reassured in early October when all the men were evacuated because the ventilation system failed and the mine gassed out. 'It was a huge relief when he came home that day. I thought, "Thank God, they've got systems to check and they've got the men's backs."' Milton, too, thought the incident was a good sign that the safety regime at the mine was working.[2]

Along with Sam Mackie and Terry Kitchin, the employees of his small subcontracting business SubTech, Osborne worked all day on November 19 installing a pipe. They stopped to eat their lunch with the McConnell Dowell workers at their underground crib-room. Riki Keane, who worked for a local building firm, Graeme Pizzato Contracting, was in the same area using a loader to prepare the floor for concreting.

Dene Murphy came off night shift at about eight a.m. on November 19; he had been on since ten the previous evening. He was shift deputy with the crew working the ABM20 continuous miner; the men were driving a roadway in the north-west corner of the mine to create a return ventilation route for a second panel that was to be mined with the hydro monitor.

Murphy had been feeling more positive about the mine over recent weeks. After struggling for so long to catch up with its ambitious promises, things at the mine finally seemed to be heading in the right direction. Working on the ABM20, which had arrived in August and was slicing through the coal with great efficiency, restored his confidence that the mine could be a success.

Underground mining always throws up frustrations, however, and there were problems with gas and equipment on his shift on the night of the 18th. The mining machine was tripping out every time methane reached 1.25 percent, but different gas detection devices were coming up with different readings. Murphy consulted with Lance McKenzie, the undermanager on duty, and together they tried to diagnose the problem. It turned out there was an in-seam borehole penetrating the floor and releasing methane into their working area. Once they located the hole they plugged it up.[3]

The day shift – B crew – came on at seven a.m. Scott Campbell was the leading hand with the group working on the ABM20. Campbell had come to work at Pike as a trainee miner in 2008, having run his own inshore fishing business in his earlier years. He was keen to develop in his new career and grabbed every opportunity to learn, including becoming a representative on the workers' health and safety committee. Despite Campbell's contribution the committee made little headway; the turnout was usually poor and the engineering department never even bothered to put forward a representative. The group regularly raised issues with various departmental managers, ranging from fire hoses left lying untidily on the ground through to the gnawing concern over the lack of a second means of egress, but their concerns were often ignored.

On the morning of the 19th, gas problems from the in-seam borehole persisted and Campbell's crew managed to cut only three metres of roadway because the machine kept tripping out. When the water supply was cut off at 12.20 p.m. mining came to a standstill, and the crew used the enforced downtime to stone-dust the area while Campbell completed his shift reports. All in all it was an unusually unproductive shift for the ABM20, which had made significant headway in recent days, including cutting 27 metres on the night shift of November 17 – three times more than the target.

At about one o'clock everyone was ordered out of the heading and into a safe area because the McConnell Dowell crew were preparing to shotfire in an area of stone they were working on. The task involved drilling holes in the face, loading them with explosives, firing the shot, mucking out the rubble, and then supporting the roof and sides with bolts and mesh.

On the short walk down to the safe zone Campbell met up with Allan Dixon, Pete O'Neill and Keith Valli and had a chat about how things were going. O'Neill mentioned they were still struggling to budge the coal using the hydro monitor and were looking at shotfiring the face to loosen it.

Campbell noticed a young visitor to the mine, too. Seventeen-year-old Joseph Dunbar stood out like a Christmas tree in his brand new reflective gear. He was due to start work the following Monday with the Valley Longwall in-seaming drilling crew, but he was so keen he had come to the mine for an orientation visit and had decided to stay on with the two-man drilling crew of Ben Rockhouse and Josh Ufer until the end of their shift.

Campbell and the other men waited for the round of explosives to be fired, but it was delayed. It was time for his shift to hand over to the incoming C crew, so they knocked off and left the mine in a driftrunner before the shotfire was completed at 1.45 p.m.[4]

At the westernmost part of the mine, close to where the Valley Longwall team was drilling, Malcolm Campbell, a young Scottish engineer, and Koos Jonker, a fitter from Limpopo in South Africa, were servicing one of the two Waratah continuous miners that was still underground.[5] The other machine had been taken to a Greymouth engineering firm for further modifications, the latest of many attempts to see if the thing could be made to work properly.

A short distance to the east, Blair Sims, a talented rugby and rugby league player, and David Hoggart were doing maintenance work near the roadheader.[6] The deputy on the previous shift, Craig Bisphan, had noted some methane layering in the roof earlier in the day; in his report the previous day he had made a stern complaint about standards. 'ABM place not stone-dusted for 15–20m of roadway focus has gone to meters and not health and safety' [sic]. On November 17 Bisphan recorded 'bad layering' of methane, with up to five percent in roof cavities, which he dispersed.[7]

The afternoon shift – C crew – were on the job a little after one, crossing paths with the outgoing B crew and then waiting in the safe zone for the shotfire before they headed to the face. Willie Joynson, the experienced Queensland miner who was so worried about safety at the mine that he and his family had decided to leave early in the New Year, was working on the AMB20. Along with him were Chris Duggan, whose brother Dan was on duty in the control room that day, Richard Holling,

Brendon Palmer, Stu Mudge, Glenn Cruse, Peter Rodger and Daniel Rockhouse, whose brother Ben was on the Valley Longwall drill rig.[8]

Dan Herk was the only deputy underground for the afternoon. Herk's father Rick Durbridge – known to all as Rowdy – also worked at Pike and had come off night shift that morning. Herk was a tough-looking character – he'd done time in prison and was an ex-member of the Lost Breed motorbike gang – but he had impressed experienced mining men with his quick mind, and his ability to pick up the mathematical and ventilation principles involved in mining. He was also diligently focused on safety, and had led the one and only walk-out of workers from the mine over a safety issue.

Other contractors were working on various projects around the mine. Michael Monk, who worked for Graeme Pizzato, and Kane Nieper and Zen Drew of Boyd Kilkelly Builders were building a ventilation stopping in an area to the west. Lyndsay Main, a Coastline Roofing builder, finished work at about 2.30 p.m. and walked out of the mine.[9]

Another building company, CYB Construction, also had employees underground. Andrew Hurren and Francis Marden were making a dirty-water sump. Another CYB Construction employee, John Hale, had become the mine's permanent 'taxi driver' and spent his time ferrying men in and out on a driftrunner.[10]

Also in the mine on that Friday afternoon were mine surveyor Callum McNaughton and a newly recruited trainee surveyor, Kevan Curtis. It was Curtis's first trip underground at Pike, although he had heard plenty about the place because his brother Matt was a fitter there, and his grandfather, Denis Smith, was a mine electrician and, at 68, Pike's oldest employee. Smith was knocking off just as Curtis was heading underground at about 12.30, and grandfather and grandson exchanged greetings and some familial ribbing.

Curtis and McNaughton had no sooner got up to the end of the stone tunnel and into the mine workings than they had to head into the safe zone and wait for the shotfire. It was a crowded place with everyone waiting together for the round of explosives to be detonated. But it gave Curtis a chance to catch up with his close friend and rugby league teammate Blair Sims.

The shotfire, when it finally occurred after almost half an hour of waiting, surprised Curtis with its intensity. 'It was huge. You're waiting and waiting – you don't know exactly when it's going to happen. Then the whole ground shook.'

After that, everyone went back to their place of work. Curtis and McNaughton finished their surveying task shortly afterwards and called on the mine communications system for a taxi ride back to the surface. While they waited, Curtis saw and chatted with Riki Keane, who by then was at Spaghetti Junction trying to restart a broken-down loader.

Conrad Adams, who was underviewer on duty that afternoon, came into the mine on a driftrunner at about 3.15 p.m., stopping to help Keane with his loader before continuing up into the mine. At about 3.30 Curtis and McNaughton managed to hitch a ride out: John Hale's taxi hadn't yet arrived, but a vehicle came by full of McConnell Dowell and Skevington Contracting workers – eight men in all – who were knocking off for the day. The McConnell Dowell men were finishing early to make up for extra time they had worked the previous day when their superintendent, Les Tredinnick, had made them stay back to fix a damaged water line. Curtis thought they seemed to be in a rush and initially reluctant to stop, but then they said, 'Yeah, jump in.'[11]

Not long after the vehicle resumed its journey down the mine they passed John Hale heading up in the driftrunner, in readiness to taxi out the men who were due to finish at four. Hale pulled to one side to let the outgoing vehicle past. By then Milton Osborne and his men would probably have been packing up and getting ready to catch a ride out with Hale's taxi, as Osborne was determined to finish at four that day.

The vehicle Curtis and McNaughton were in also passed miner Russell Smith driving a loader up the stone tunnel towards the mine workings. Smith was running late for work: he had forgotten that, because it was a Friday, his shift started an hour earlier than usual.

At about the same time, Daniel Rockhouse was driving a loader down from the ABM20 place in the north-west corner of the mine towards pit bottom in stone, an area adjacent to the stone tunnel which housed pumps, water and electrical services. He was going to collect some gravel to fix a roadway.

Curtis, McNaughton and the eight McConnell Dowell and Skevington men left the tunnel portal at 3.42 and headed to the mine offices to finish up for the day.

They heard nothing at all when the mine exploded at 3.45.

The only witness was a camera located at the portal: it recorded the sudden movement of a telltale rag hanging at the entrance, and then 52 seconds of debris-laden air moving with furious force out the mouth of the 2.3-kilometre tunnel.

In the surface control room, Dan Duggan was talking to Malcolm Campbell on the underground communications system; he was turning the water back on and needed to let the men underground know. The conversation suddenly went dead, and alarms went off in the control room. Duggan kept speaking into the machine, trying to raise a response from somewhere – anywhere – in the mine, but no one was answering.

Doug White was in his office, having a meeting with Steve Ellis and George Mason. The three of them were discussing whether shotfiring should be used to help ease the hard coal that was proving so difficult to budge with the hydro monitor. White was opposed to the idea; the hydro system was still being trialled, and they were trying to figure out the ideal water pressure and angle of cut; he thought it would be a mistake to change more than one parameter at a time during this phase.[12]

At 3.45 the lights flickered, but the men thought little of this: it wasn't an unusual occurrence. A few minutes later Duggan called White to tell him that power was out and communication with the men underground had been lost. This didn't alarm White either, as it too had happened before. A few minutes later Duggan asked White whether he should call the Mines Rescue Service. White said, 'Oh, we won't go there yet. We'll get someone up there.' The computer screens in the control room were flashing red, indicating a fault. White went outside and noticed a strange smell in the air, like diesel fumes or something similar. The mine portal was out of sight, about one kilometre further up the road. It didn't occur to him there had been an explosion.

He went into his office and wrote three emails in connection with job possibilities that would enable him to leave Pike. Peter Whittall's accusation earlier in the week that he had caused a seven-cent drop in the company's share price by speaking unguardedly to financial analysts had been the final straw. He was determined to leave.

Soon after the meeting with White and Mason ended, Ellis headed down the valley to Greymouth. He had to get to the post office before it closed to send a registered letter relating to the sale of his house in Australia; he was buying a home in Paroa, just south of Greymouth, for him and his family.[13]

White went up to the portal and spoke to engineering manager Robb Ridl and electrician John Heads. Nothing appeared to be amiss at the entrance to the tunnel, just a piece of brattice cloth lying on the ground. There was no smell, and ventilation was entering the tunnel.[14] Another electrician, Mattheus Strydom, had been instructed by Ridl to go into the mine and investigate the loss of power.

Strydom advanced some way up the tunnel, feeling increasingly uneasy and finding it harder and harder to breath. Then he saw Russell Smith lying on the roadway, as if dead. Strydom was afraid of losing consciousness himself. He quickly reversed down the tunnel until he was able to turn around at a stub, and headed straight to the portal. When he got there he phoned the control room and told Dan Duggan, 'You'd better call the Mines Rescue. We've had an explosion and I've seen a man lying on his back in the road.'

It was now 4.25 p.m. Duggan, whose instincts had been telling him for half an hour that the situation was bad but who had deferred to

White's judgement that it was not yet necessary to alert the Mines Rescue Service, was finally able to call for help.

At about 4.50 Doug White called Steve Ellis on his cell phone: 'There's been a big bang here, Steve. It's serious. Can you get back to the mine?' Ellis replied, 'I'll be there directly.'[15]

By now Daniel Rockhouse had regained consciousness where he lay on the ground at pit bottom in stone; he had been knocked over by the force of the blast and then poisoned with carbon monoxide fumes. Having willed himself to get to his feet and move, he had found his way through the blackness to the compressed air line and opened a valve to flush his face and lungs with fresh air. At 4.40 he managed to communicate through an underground phone with the control room. He then began his long struggle out of the mine, gathering up Russell Smith from where he was crouched, semi-conscious, near his loader in the tunnel. Rockhouse hauled himself and Smith to safety – a feat of endurance that ended when they emerged from the portal at 5.26 p.m.

Sergeant Dave Cross had not long come on duty at the Greymouth Police Station when the emergency call came through. He headed immediately to the mine with Constable Shane Thomson, arranged to have a cordon placed across the mine road, and carried on up to the administration offices. The Mines Rescue Service, the highly specialised group of brigadesmen trained to respond to mine emergencies, had not yet arrived. Cross was initially told there could be as many as 36 people underground – the numbers were uncertain because the tag board, where workers hung their name tags, was not always accurate. Sometimes men forgot to take off their tags at the end of a shift, and sometimes they forgot to tag on when they went underground.

Greymouth mayor Tony Kokshoorn, a man who had championed and supported the Pike River mine for years, was also on his way. He had been in his car going to pick up his wife, Lynne – the couple planned to go up to their bach at Lake Brunner and look at a new lounge suite that had just been delivered – when three ambulances screamed past. He thought perhaps there had been a bus accident on the coast road. When he had arrived home, he had received a call from one of his councillors, Peter Haddock, who had word of an explosion at the mine. Immediately after he hung up, the police called to say there had been an 'accident' at Pike River, with between 25 and 30 men missing. Kokshoorn sat, momentarily paralysed, and then said to his wife, 'The fucking thing's blown.' Lynne chided him for his language. 'No,' he said, 'it's Pike. It's blown its top.' He took off 'like a nutter. I felt I just had to be there.'[16]

He drove at 150 kilometres an hour and faster up the Grey Valley to the mine. By then, news of an explosion had begun leaking out. Cell-phone reception up the valley was patchy, but some calls got through to Kokshoorn as he tore around bends at high speed. Among others, the BBC was on the line wanting to know the latest. When he reached the Pike miners' bathhouse at 5.30 Anna Osborne was already there, as were Tom and Debbie Daly. Tom Daly was the chief of Ngāhere's volunteer fire brigade and Milton Osborne was his deputy, best mate and whitebaiting buddy. They were all desperate for word of him and refused to budge when mine staff tried to make them leave.[17]

When Kokshoorn got to the mine administration buildings there was little happening. Doug White had gone by chopper to inspect the back-up fan at the top of the ventilation shaft. The fan was supposed to kick in if the underground fan stopped working. It turned out to be badly damaged. So too was the adjacent communications shed, through which data from mine gas sensors, the underground fan and pumps was normally marshalled and sent to the control room. The generators that drove the back-up fan had also been hit by the force of the explosion and the surrounding bush was blackened.

The atmospheric conditions inside the mine were unknown – the real-time gas monitoring system, such as it was, had been destroyed. Had there been in place the tube bundle gas-monitoring system that White wanted to install at Pike, and which would have been powered from the surface, it may have been able to continue drawing out samples after the explosion. As it was there was no information at all.

Sergeant Dave Cross was meanwhile trying to create order amid chaos. He had never been to the mine before. He didn't know that the portal was a further kilometre up the winding road from the administration buildings, nor that the mine workings were more than two kilometres uphill through the mountain. He was, though, familiar with the role of the Mines Rescue Service and knew they would be the men who would undertake any rescue attempt. He saw the job of the police as providing incident management, fronting the media, and bringing their considerable resources to bear on the operation.

It was almost seven o'clock before Cross was given a list of names of the 29 men who were underground, plus another one whose whereabouts was uncertain. By then, news of disaster had erupted across the nation. There was word that two men had escaped. An unfounded story was circulating that three more had been seen making their way out. Those who knew little of the chemistry of coal and the tragic history of underground coal mining began wishing for a heroic rescue like that which had recently occurred in Chile, when 33 men trapped underground in a gold and copper mine for 69 days had been brought up alive.

Cross and his fellow officers were, however, looking at every eventuality. By 7.30 they were sourcing body bags. It was almost four hours since the explosion and no one else had come out.[18]

Rob Smith and Troy Stewart, safety and training officers employed by the New Zealand Mines Rescue Service and based near the settlement of Rapahoe, had been at work since five that morning, running an emergency response training course at the Stockton opencast mine north-west of Westport. They were driving back to Rapahoe at about 4.45 in the afternoon when they received a phone call on one of the few stretches of the scenic coast road with cell-phone coverage. Glennis Lemon, the administration officer at the rescue station, had news of an incident at Pike.

They drove on, Smith in the passenger seat going through operational procedures, and Stewart driving in as much haste as he could amid the busy tourist season traffic of camper vans and rental cars. By the time they got to the rescue station, volunteer brigadesmen had already started assembling themselves and their equipment and were waiting to travel up the valley to Pike.

The general manager of the Mines Rescue Service, Trevor Watts, was away in Huntly on business that day, and had been briefed by phone. Watts instructed Smith, who took charge in his boss's absence, that it was essential no one went underground until there was adequate information about the atmospheric conditions; the safety of mines rescue brigadesmen was of utmost importance.[19]

The chopper flew Smith and his crew over the ventilation shaft before heading to the mine administration offices. It was a sobering scene. Misty smoke was rising from the shaft, and there was a great deal of soot over the surface fan and extensive damage to surface structures. It was clear to Smith that the situation was grave. The smoke indicated fire, which meant the atmosphere would be full of asphyxiating carbon monoxide, and it was obvious from the badly damaged structures that the explosion had been large.

They had time to fly around the shaft four times while they radioed through to Pike to ask for the yard to be cleared so it could serve as an emergency helipad; in the end they had to put down in the riverbed of Pike Stream and haul their gear on to a waiting minibus.

It was close to six when they got to the mine office. Things seemed to Smith to be chaotic, with ambulance, fire service, police and mine staff in attendance and no obvious sign of a formal incident management process having been established. Daniel Rockhouse and Russell Smith had not long emerged from the mine and were being loaded into an ambulance. They were both black with filth and had red glassy eyes. It looked to Smith as if they had inhaled a 'fair gutsful' of carbon monoxide and were not well. They were able to walk with support to the ambulance but were in no state to talk.

Aside from what they had seen of the two survivors and the damage at the top of the ventilation shaft, the brigadesmen had no other information upon which to assess the situation. Smith and Doug White both knew they had to somehow get atmospheric samples from the mine. Until then there could be no attempted rescue. White, who considered he had control of the site as the statutory mine manager, barred anyone from entering the mine until air quality readings had been obtained. Smith couldn't disagree. 'We don't go rushing into mines. Mine Rescue also has a duty of care under the Health and Safety in Employment Act to its employees and brigadesmen. We needed that information before we went anywhere. It was incredibly frustrating that we didn't have it,' he would later recall.[20]

The only way to get samples was to fly up to the top of ventilation shaft and collect them from there. Mine's Rescue brigadesmen Rick Roberts, Richard 'Junior' Banks and Russell Smith (whose namesake and fellow Pike worker had just emerged alive from the mine) were dispatched to do this. They also tried to establish communication with any survivors underground by lowering a bucket containing water and a radio phone down the slimline shaft, the 600-millimetre borehole that had been drilled in 2009 to augment ventilation after the main shaft had collapsed. There was no response. They managed to collect a few air samples, which were ferried away by chopper. Then cloud closed in, preventing the helicopter from lifting the men off the mountain. They were forced to spend a cold night on the tops, not knowing if the mine directly beneath them would blow again.

Smith and his team were desperately frustrated. They had assembled their gear and had trained for this very situation and wanted to rush in, but couldn't. At one point brigadesman Mike Nolan said to Tony Kokshoorn, 'We don't give a damn about the gas, we just want to get down there.'[21]

As part of their training they had studied coal mine disasters in their own region and around the world, and some believed there was a 'window of opportunity' for rescue immediately after an explosion when all the available methane has been consumed by the blast. But it was also known that Pike was a very gassy mine and would fill with methane in a matter of hours if ventilation stopped; there had been clear evidence of that just a few weeks earlier, when the mine gassed out on October 6 after the surface fan failed, and everyone had to be evacuated.

Smith also knew that secondary explosions were common, as a build-up of flammable gases was ignited by the burning residue of the first blast. This had happened many times in mines. In some cases the second explosion would follow only minutes or hours after the first; in some cases it would be days later. Just six months earlier an explosion at the Raspadskaya mine in Western Siberia had been followed by a second blast four hours later, killing 19 rescuers.

A further complication was that the only way in and out of the mine was on foot via the 2.3-kilometre uphill tunnel. It would be like walking up the barrel of a gun; if there were another explosion there would be nowhere to hide.

All that Smith's team knew of the conditions underground at Pike that evening was that methane and oxygen were present, and judging from the smoke emitting from the shaft there was possibly fire. As fiercely as they wanted to mount a rescue effort – the men underground were their neighbours, friends, workmates, relatives – to go in without more precise atmospheric information would be a reckless gamble. Any re-entry had to be managed and motivated by rational heads, not grieving hearts.

Hours had now passed and no more survivors had come out of the mine. Smith was increasingly sure the situation was catastrophic. Later in the evening that sense of foreboding deepened when co-worker Troy Stewart and brigadesman Chris Menzies went to the control room and saw for the first time the video footage taken at the portal at 3.45. The footage provided the men with vivid evidence that they would not be rescuing men alive; their job would be to recover bodies. They had both been at the mine on training exercises and knew it to be very small – it had barely scratched the coal seam – but the explosion had gone on and on for 52 violent seconds. There would have been nowhere in the tiny grid of roadways for men to escape the ferocity and toxic aftermath of such a blast. It was a moment of painful despair: of the men underground, five were Stewart's friends.[22]

The Fire Service, who were assisting at the scene, were also pessimistic. At ten p.m. a note in the service's incident log recorded that atmospheric readings showed 700 parts per million of carbon monoxide. A further note 19 minutes later recorded: 'Confirmed that at 600 ppm [CO] for 30 minutes, expect fatal.'[23]

By then, however, the highly trained Mines Rescue brigadesmen and local police were being quietly stripped of control. Pike was shaping up as an incident of national importance, and an 'O' desk had been created in Wellington. By the time Sergeant Dave Cross left the mine at midnight, he had the 'strong impression that power had left the site'.[24]

Tony Kokshoorn had meanwhile been thinking of the families of the missing men. What had they been told? Where could they meet to be kept informed? What support would they need? He made arrangements for the Red Cross building in Greymouth to be set up as a welfare centre, and told Doug White he was going to head back to town to speak to the families. Kokshoorn recalls White resisting and saying: 'We [Pike] will front this.'

By 10.30 p.m. Kokshoorn was still waiting for someone from the company to come to town to speak to the families; they had been reliant for information on news broadcasts and the West Coast grapevine, and each family was desperately searching for confirmation that their family member was safe, or missing. At about eleven Kokshoorn entered the Red Cross building. Only about 15 people were there. Other families were still making their way through the night from Christchurch, Southland and beyond. Many had no idea where to go for news, having tried in vain to find out something from the company and the police. Some had tried to drive to the mine site, only to be turned away at the police cordon.

Word that two unnamed men had walked out excited hope among some that it could be their son, brother, husband or father who was safe. The rumour – baseless though it was – that three more had been seen making their way out of the mine allowed optimism to survive the night.

But those who came from coal mining families, or who had mining friends willing to tell them the brutal truth, already knew there would be no repeat of the sensational Chilean rescue. Those 33 men had survived because they were in a gold and copper mine, with no noxious gases. Long-term survival in the poisonous aftermath of an underground coal mine explosion would likely be impossible.

The Red Cross room was thick with anxiety. Kokshoorn told the group what he knew of events at the mine, and of the need to gather information on gas levels before there could be any attempted rescue. None of the families in the room had yet been contacted by the company. The office of the prime minister, John Key, had however received a call from Pike's board chair John Dow earlier in the evening to create a point of contact and advise that the company was taking the event very seriously.[25]

Kokshoorn was enraged at the lack of consideration for the families' need for information. He gathered up their phone numbers, stuffed them into his pocket on scraps of paper, and sped back up to the mine. When he arrived he had a stand-up confrontation with Pike's human resources manager Dick Knapp, who Kokshoorn had understood was responsible for informing the families.

At 2.40 a.m. Peter Whittall, the man who had driven the Pike development for almost six years and who had recently become its chief executive, arrived from Wellington. Whittall had been busy dealing with matters relating to the upcoming $70 million capital-raising when he had received a call at 4.45 p.m. from Robb Ridl to say there had been an explosion in the mine. He had asked Ridl if it were an emergency drill. When he realised it wasn't, he felt numb. Pike had only five or six head office staff but that day there were also sharebrokers on the premises talking to Pike's financial controller, Angela Horne, about the upcoming equity raising. Whittall stood shaken and alone in his office.

Within minutes he was inundated with calls from the media, and from business people who had heard the news via Twitter or news reports. Television crews turned up at the office wanting interviews.

Offers of assistance poured in, including from Air New Zealand chief executive Rob Fyfe, who suggested he use the airline's highly trained crisis response team.[26]

Whittall left the head office at about 8.30 p.m., caught the last flight to Christchurch, and then drove over the mountains to the mine. He was accompanied by Miranda Clayton of Busby Ramshaw Grice; Pike had approached the public relations firm at about 6.30 to assist with communications through the crisis.[27]

Whittall was disappointed to discover on his arrival that little progress had been made in contacting the families. With his endorsement, Kokshoorn helped sort through the incomplete, and in many cases inaccurate, list of next-of-kin contacts held by the company, and contributed the numbers he had collected from the meeting at the Red Cross centre. Finally, as dawn approached, the company began making contact with some of the families of the men underground.

Many never received a call.

ELEVEN
Five Days

The phone rang before dawn on Saturday, November 20 at the Barrytown home that Lauryn Marden shared with her husband Francis and their young children. Francis worked for CYB Construction, which subcontracted to Pike River Coal. He was 41 and had been at Pike for 16 months. He was a practical man who could turn his hand to virtually every aspect of the building trade – he and Lauryn had built two houses together – and had accumulated the certificates needed to operate several of the mine's machines. Although he wasn't happy with the conditions and long hours at Pike, he was pleased with what he had achieved there and was willing to persevere.

Lauryn had learned of the explosion at the mine at about 4.45 on Friday afternoon, when Francis's supervisor called to say he had heard the news over the police scanners. She had tried in vain to get through to Pike River's control room, but no one had answered. She tried the police but they couldn't tell her anything. She fielded calls from worried family members and friends who had seen the news on TV, and hoped desperately Francis was somewhere in transit from work and would show up at home soon.

He didn't arrive. At midnight Lauryn was taken by Karen Greenslade, the wife of Francis's boss, to Pike's small Greymouth office, but she learned nothing there. She went home and waited. When the phone rang at five a.m. the woman at the end of the line simply told her to be at a meeting in Greymouth at seven. She asked if this was the call she had been expecting from Pike, advising her that her husband was down the mine. The caller said 'Yes' and that was the end of the conversation.[1]

Tara Kennedy got the call at 5.30 a.m. Her husband, Terry Kitchin, worked for Milton Osborne's firm SubTech. Like Francis Marden, he was 41. Tara had been called by her father at five the previous evening after he heard news of an explosion at Pike on the radio. Along with two friends, she had decided to drive to the mine. She didn't know where to go at first: Terry had been working there for only three months.

On the way, the radio had reported that an information base had been established at the Moonlight Hall up the Grey Valley, and also at the Karoro Learning Centre in Greymouth. The three women headed for the Moonlight Hall but there was nothing there, not even a light on. They attempted to drive onwards to the mine but were barred from going beyond the cordon. Along the way they ran into Civil Defence staff, who told them they should go to the Red Cross centre. Tara got home from the centre at about three a.m. and waited. The 5.30 phone call told her to go to a meeting at the Red Cross centre at seven.[2]

At the centre, family members were jammed into the small room, lining the walls and sitting on others' laps. There was scant news. Pike River chief executive Peter Whittall and Greymouth mayor Tony Kokshoorn spoke.

Police superintendent Gary Knowles was also present. A 33-year veteran of the police force, Knowles had been on leave at his Nelson home when he had been phoned the previous afternoon by his superiors and told to hasten to Greymouth and take command of the situation as incident controller. His trip to the West Coast had taken longer than normal. He had had to detour to pick up Police Tasman District communications manager Barbara Dunn from her rural home. After 45 minutes, the pair received a request from the police communications centre to collect 36 body bags from Nelson and so they had had to turn back. When they arrived at Greymouth they didn't know the way to the mine; with no cell-phone coverage as they travelled down the Grey Valley, they had had to pull over a milk-tanker driver to ask for directions. They finally reached the scene at 12.20 in the morning.

Knowles told the families that safety would be the paramount consideration in the rescue operation. Whittall reassured them that a compressed air line was running into the mine and drinking water was available for the men. The message was not to worry.[3]

The fast-growing contingent of national and international media was briefed next. The line was cautious but optimistic: the *Greymouth Star* reported that it could be 'tea time' that day before a rescue would be undertaken[4] but that the men were hopefully waiting at the fresh air base. Journalists observed that Whittall was visibly tearful, and reported him saying, 'I've been there for every metre cut. It's very personal.'[5]

Only those intimate with the mine would have known that there was no proper fresh air base, merely a small working area that had been given the name. Cut into a mine roadway, it had a rough floor, bolted and meshed sides, and a roll-down brattice cloth door. An overpressured methane drainage line ran through it, and it was too small to accommodate all the men who might be underground in an emergency.

Throughout the morning, reports swirled that the two survivors had got out via an escape route up the ventilation shaft, emerging unhurt. Reporters who were ignorant of the real facts about Pike River mine could not have known how preposterous this suggestion was. Escape up the vertical ventilation shaft in a toxic and oxygen-depleted atmosphere would have been impossible, and was known to be so by everyone in the mine.

Nowhere – neither at the family meeting nor the media conference – was there mention of the enormous force of the explosion, or that there was video footage of the blast exiting the portal, footage that had been viewed the previous evening by several people from the Mines Rescue Service, Pike River Coal and the police.

While Whittall spoke of his hope that the men would be brought out alive, Robin Hughes was at the Rapahoe Mines Rescue station analysing the atmospheric samples that had been gathered with great difficulty from the ventilation shaft.

Hughes was an experienced mine ventilation engineer who had worked for 40 years in the coal mining industry, including four as chief inspector of coal mines. The lean and athletic 61 year old had been out biking when his wife had called him on his cell phone with news that the police were looking for him because there had been an explosion. He had rushed to the station at Rapahoe and stayed there until midnight, helping set up a roster of brigadesmen, and preparing the station's gas chromatograph to analyse the samples that would be flown down from the mine site. He managed to snatch some sleep at home before reporting in again at six the next morning.

The first samples that arrived at the station were diluted by fresh air and unreliable. Getting quality samples out of the ventilation shaft was difficult and dangerous: if the mine exploded again and men were at the top of the shaft they could be killed. An ingenious solution was found. A stomach pump from one of the ambulances waiting at the mine was hooked up to a long piece of tubing, flown up the mountain by chopper, and lowered down the shaft. Not only was this method safer, it enabled better quality samples to be collected from deeper into the shaft.

The samples were flown down to the Rapahoe station in four-litre anodised plastic bags. There, using the gas chromatograph to identify the constituent gases in each sample, and then applying several gas ratios to calculate the temperature, degree of oxidation and products of combustion, Hughes began to assemble a picture of the likely conditions underground.

An equation called Graham's Ratio measures the efficiency of the conversion of oxygen to carbon monoxide, and is used to determine the presence of heatings or fires in coal mines. Ratios of less than 0.4 may be found in normal mine air or goafs. A ratio in excess of 0.4 indicates the possibility of combustion in the coal. A ratio of 2 suggests that fire is almost a certainty.

The samples from Pike were 34, off the scale. Hughes ran the results through a second test, the Jones-Trickett Ratio. It supported his conclusion that a methane fire was burning underground.[6]

Mines Rescue Service general manager Trevor Watts, who had travelled through the night from Huntly and reached the mine at three in the morning, arrived at the station at about midday. Hughes said to him, 'The mine's on fire, Trev.'

By then Watts was one of the many who had seen the CCTV footage in Pike's control room. He had applied his understanding of the mechanics of explosions, and immediately understood that no one could have survived the force of this blast in such a small mine.[7]

Not only this, but all the ingredients were present for a second explosion: an ignition source, flammable gas and oxygen. Unless the mine atmosphere could be brought under control by blocking the flow of oxygen, it was a matter of when, not if, it would blow up again.

Hughes travelled up to the mine and attended the on-site incident management meeting at three in the afternoon. A large group had assembled – police, fire brigade, ambulance staff, Red Cross workers, mines rescue men, Pike managers, and Department of Labour inspectors. Steve Ellis, Pike's production manager, was chairing the meeting as the company's most senior official at the scene. By then, he and the statutory mine manager, Doug White, were sharing the responsibility: Ellis was doing the 12-hour day shifts and White was working through the night.

Hughes told the group of his conclusion that there was almost certainly a fire burning underground, and that he had been getting the same results consistently since ten that morning. As he spoke he looked across the room and saw shock pass across the face of local policewoman, Senior Sergeant Ali Ealem.

Craig Smith, who worked for Solid Energy as the company's underground mining manager, was also at the meeting. Like Hughes, Smith was a member of the Mines Rescue Trust board and had four decades of experience in the industry. He had no doubt as to the significance of Hughes' analysis, and was worried the brigadesmen might come under undue pressure to enter the mine, when it was clear the place could explode again at any time. He was sure that after almost 24 hours with no contact from anyone underground, no one was alive. The immediate challenge was to stabilise the atmosphere to maximise the chance that the mine could be re-entered and the men's bodies recovered.

But in front of the crowded meeting – where the majority of those in attendance were laypeople – Ellis cast doubt on Hughes' gas results. The readings, he said, were probably only indicative of fumes from the explosion.[8] Hughes was furious. After he and Smith left the meeting the two men had barely walked a couple of steps before they turned to each other and said, 'We've got to seal the mine – now.'[9]

The Fire Service, too, had formed a bleak view of the situation. At 7.10 a.m. on Saturday, November 20, it was noted in the service's incident log: 'Planning needs to commence for mass fatality (not public) … concerns that the mines company do not fully appreciate the gas levels.'

At 9.36 a.m: 'Situation remains unchanged. Anticipate K41 [Fire Service radio code for a fatality].'

At 3.21 p.m. the log noted: 'Anticipate that there is a fire burning.'

At 4.45 p.m., 25 hours after the explosion: 'Local staff are reporting significant fire underground. What can NZFS do to assist?'

At 5.19 p.m: 'All indicators are positive that it's a fire. Options to seal and fill with nitrogen about the only way you could deal with this kind of fire. … 3 fires on the go in other mines that have been burning since the 1950s.'

At 9.12 p.m: 'Need to start advising families as to what's happening.'[10]

The families were not advised. On Saturday afternoon they were called back to the second meeting of the day, setting in place what would become the anguished routine for the next four days. There would be a meeting at about eight in the morning, at which Pike chief executive Peter Whittall and police superintendent Gary Knowles, the incident controller, would speak. This would be followed by a media conference at ten. At about 4.30 in the afternoon the pattern would be repeated.

When John Key visited from Auckland that day he was given the same hopeful message as the families. 'So they could be alive?' he asked Whittall. '"Yes," Whittall said, "they could be alive – some could be dead and some could be alive. It's a big space down there and it depends where they were in the explosion, and these two guys had walked out and it's a solid structure and it depends where the explosion took place…"'[11]

Like most other New Zealanders and many of the missing men's families, Key didn't know much about coal mining, or about Pike River Coal, other than that the company had 'talked a pretty big game'.[12] Gerry Brownlee, the minister of energy and resources and ranked number three in Key's Cabinet, had cut the ceremonial ribbon at the official opening of the mine two years earlier. And, like others, Key had in mind the images transmitted around the globe five weeks earlier: 33 Chilean gold miners being lifted to safety after being trapped deep underground for more than two months.

Sunday brought little change. Robin Hughes continued to assist in the analysis of gas samples at Rapahoe, and the results continued to indicate fire. Parallel analysis done on another gas chromatograph that had been flown to the mine by Queensland's Safety in Mines Testing and Research Station (SIMTARS) confirmed the Rapahoe results.

The New Zealand Mines Rescue Service had by then mobilised some of the most experienced mining men in Australasia to the site. At yet another crowded incident management meeting in the early afternoon, the consequences of failing to take control of the mine atmosphere by stopping the flow of oxygen were spelled out. Hughes explained that unless the existing fire was controlled, it would become more extensive. An intense fire in an underground environment would inevitably destroy the integrity of the roof support, causing cave-ins. If that happened, recovery of the men's bodies might become impossible.

At the meeting at six that night, the Mines Rescue men formally proposed that the mine be sealed in the hope this would reduce the chance of a second explosion. Not all the assembled experts agreed.

Darren Brady, a gas chemist from SIMTARS, thought blocking the flow of oxygen could actually precipitate a second explosion by changing the gas composition and ventilation inside the mine. He also recognised it would be a brave call for anyone to seal the mine if there were even a sliver of chance that someone might still be alive.[13]

The matter was far from clear-cut, but even if there was to be no immediate decision to seal the mine, at the very least there needed to be robust discussion of the options. And parallel planning needed to get underway so that sealing could proceed quickly if it was determined to be the best course of action. There was only one certainty: if nothing were done, the mine would blow up again.

The proposal to seal met a dead hand and discussion was shut down.[14] Representatives of the Department of Labour – the same organisation that had considered Pike a compliant, best-practice organisation and turned a blind eye to its multiple breaches of safety regulations over the previous two years – made it clear that any move to seal the mine would be blocked. Until there was evidence of 'zero life' underground, the focus would remain on rescuing survivors.

Yet the overwhelming evidence was that there could be no one alive. Carbon monoxide levels were such that a person without breathing apparatus could not survive. The self-rescuers carried by the men provided 30 minutes of air, and even if the men had made use of the cache of spare self-rescuers, the life support these afforded would have long since expired. There was no place of refuge in the mine where men could seal themselves off from the noxious atmosphere and breathe clean air. There had been strong evidence of a fire burning since Saturday morning. Hourly calls had been made over the mine's underground communication system and no one had answered. The men had been trained to walk out in an emergency if they possibly could, but no one had emerged in the 48 hours since the miraculous escape of Daniel Rockhouse and Russell Smith.

Nonetheless the mirage of hope was maintained. Ellis had seen the CCTV footage of the explosion on the Friday night and had been shown the evidence of fire, yet he maintained some men could still be alive.[15] In private, however, he expressed a different view. The weekend after the explosion he visited Daniel Rockhouse, who was struggling in the aftermath of his ordeal and grieving for his missing brother and 28 other workmates. Referring to the handbook of the New South Wales Mines Rescue Service, Ellis told Rockhouse that the shock wave from the first explosion would have killed all the men further into the mine. The young survivor was struggling with feelings of immense guilt, and Ellis was trying to assure him: 'It's not your fault, son.'[16]

Away from the site – in Greymouth, where the families trudged to the twice daily meetings, and across a tiny nation that held its collective breath for good news – Peter Whittall became the link to the mine and to the missing men. Among those who lacked mining knowledge or access to informal reports from those at the scene, his compassionate manner and hopeful message that the men could be sitting somewhere sucking on a fresh air pipe was deeply reassuring. Tara Kennedy would go home to the couple's three children – two of whom shared the birthdate of November 25 – and tell them 'daddy would be home in time for their birthdays'.[17]

Kath Monk, too, hung on to every optimistic word in desperate hope that her 23-year-old son Michael, a young carpenter with a luminous smile, would come home. Her husband Bernie knew better. A mining friend, Dave Homson, had visited him at home late on the Friday night and told him that in his heart he hoped the men would get out, but 'talking to me straight, as a friend, he had said, "Michael is not getting out alive."'[18]

Bernie told Kath what Homson had said, but she would not hear it or believe it. Michael was young, fit and strong. She thought that if anyone could get out, he would. At the family meetings she hung on Whittall's words.

Kath Monk visited the mine site on the Sunday, part of an organised bus trip for family members. Like most of them she had never been there before and was struck by the beauty and peace of the place. She took a stone and banged it on the water and coal slurry pipes that ran from deep in the mine to the coal processing plant ten kilometres away; she felt it was a way of sending a message to her son that he must keep up his strength, that his family was waiting for him.[19]

Other family members who had friends and relatives working at the mine site were incredulous at the gulf between what their contacts were telling them and what they were hearing at the family meetings. Carol and Stephen Rose, the mother and stepfather of a missing miner, Stu Mudge, were visited by a mining friend on the Saturday night. He told them the mine was on fire and that no one would be coming out alive. They went to the family meeting early on Sunday morning expecting to be told the rescue operation was all over. Instead Peter Whittall walked into the hall with a mine map under his arm and proceeded to tell the families the men could be at one of the fresh air bases and would be hungry when they came out. He spoke to the families of a 'heating' akin to 'smouldering rags' or 'a gas hob burning in a kitchen' rather than fire.[20]

Journalists were also hearing rumours that the mine was on fire, and anyone in Greymouth with strong mining connections knew of it. Former mining journalist Gerry Morris, whose brother-in-law and two great-nephews worked at Pike, arrived from Wellington at about eleven on Sunday morning. Within a couple of hours he picked up word of the fire from miners already grieving for their lost mates.

Morris went to the afternoon press conference that day expecting – just as Carol Rose had done a few hours earlier – to be told the terrible news. Instead there was more talk of smouldering and 'smoky vapour', and assurance that fresh air was being pumped into the mine. Morris, who had grown up in Greymouth and knew some of the missing men's families, was outraged. Tony Kokshoorn, an old Marist School and rugby league friend, was also at the briefing. Morris shook his finger at him and hissed, 'Koko, they're fucking dudding you. The mine's on fire.'[21]

Tensions were already running high at the press conference, with Australian journalist Ean Higgins labelling Gary Knowles a 'country cop' and asking why he was running the operation. At the end of the conference, Tony Kokshoorn demanded that Brian Small, one of Pike's media advisers from Busby Ramshaw Grice, bring Peter Whittall back to answer questions about the fire. Whittall was unavailable.

The Fire Service continued to hold a grim view of survival prospects, although the families didn't hear this. At 8.49 on Sunday morning a note was recorded in the service's confidential incident log for Operation Pike: it had been 'flagged to Police that there is a need to tell the families given the length of time we have collectively known of the fire'. At 3.30 p.m. that day the log recorded: 'CO_2 readings at 9.6% indicate a probable internal temperature of 4500 degrees [centigrade]. Graham Combustion scale of 34.28.'[22]

They were not the only ones who knew the truth. At 6.30 on Sunday night, a Police National Headquarters situation report noted that gas samples indicated 'huge and significant combustion. Potential for secondary explosion, greater than the initial explosion ... Mine rescue developing options to stop the fire – this may involve sealing off the mine (starving it of oxygen) – there is a significant risk of secondary explosion whether the mine is sealed off or not. No consideration of any attempt to quell the fire will be considered until there is a 0% risk of survivability.'[23]

Peter Whittall, who had always insisted on being kept closely informed of operational matters at the mine, even after he moved to Wellington in early 2010, had set himself up at the Coleraine Motel in Greymouth's High Street, and was receiving morning and evening briefings about the rescue effort from Steve Ellis and Doug White. While frantically busy, he was closely supported by Nick Thompson of the private security and intelligence company Thompson & Clark, and by his capable and well-organised personal assistant Catriona Bayliss. He did not seek out detailed briefings from the many other experienced men at the scene, such as SIMTAR's Darren Brady, Mines Rescue general manager Trevor Watts, or Robin Hughes, who had been analysing gas samples since Saturday morning. He considered that as chief executive he ought to leave the operation to those at the site and focus on managing communications with the families and the media.[24]

He also met regularly with Gary Knowles. The police superintendent, too, had decided he ought to be at an emotional distance from the scene; he had based himself at Greymouth, although he visited the mine every day and received situation reports from his frontline staff at least hourly. It had been made clear to him that critical decisions – such as permitting the Mines Rescue team to re-enter the mine, or moving to starve it of oxygen – would be made by his superiors at police headquarters in Wellington. Knowles' role was to keep the families up to date and corral the resources of the police to support the operation – including tasks such as getting a tube bundle gas monitoring unit brought into the country, and arranging for a track to be cut up the mountainside to the ventilation shaft.

Knowles was told on the Saturday that there was evidence of a fire burning underground.[25] But he had also been told there was a 'clean room' in the mine, that fresh air was flowing in, and that some of the experts at the scene thought some men could still be alive. Whatever the chance of this, he was determined no further lives would be risked by allowing rescue teams to enter the mine while atmospheric conditions remained volatile and unstable.

By the Sunday afternoon, numbers at the family meetings had swelled to 400 or more, and the venue had been shifted to the Greymouth Civic Centre's indoor basketball court. The place was a chamber of despair, confusion and shattering disappointments. Morning and night, Knowles had to stand up and tell the assembled relatives that the conditions underground were still too dangerous to attempt to save their men. As the hours and days wore on, the grief and fury of many of the families turned on the police officer, who came across as wooden and lacking in empathy, while Whittall impressed as the bearer of hope, the brave and caring face of the crisis.

Up at the mine site, those tasked with assessing the conditions underground and mounting any re-entry effort had been rendered largely impotent. Sergeant Dave Cross's hunch late on the Friday night that power had left the site was correct: the decisions were being made hundreds of kilometres away in Wellington.

Seven of New Zealand's 13 most highly qualified mine managers were at the scene, but critical tasks – such as drilling additional boreholes and putting a camera down the slimline shaft – were being governed by a convoluted three-tier risk-assessment process. Documents setting out the risks and how they would be managed would be prepared by the experts at the mine, then emailed to Knowles in Greymouth for review. Knowles would send them to the local Department of Labour officials for comment, and then on to Assistant Commissioner Grant Nicholls at Police National Headquarters, from where they would be flicked across to officials at the Department of Labour's head office. They would then be sent back down the line to the mine – sometimes unapproved.

None of the people in this labyrinthine chain of command knew anything about coal mining. Indeed Nicholls, who was making the final calls, searched 'Mines Rescue Service' on Google for information as to what it was.[26] Members of the police team in Greymouth, who had assembled from other parts of New Zealand and colonised the desks and offices of the local constables, were subjected to 'Mining 101' lessons from two Department of Labour inspectors, one of whom had no background in mining.

By Monday, neither Whittall nor Knowles had yet seen the most vivid piece of evidence about the explosion – the segment of footage from the CCTV camera at the portal that showed the explosion erupting out of the mine at 3.45 on the Friday afternoon. Several Mines Rescue brigadesmen, Pike workers and police officers at the site had seen it; Whittall and Knowles, the two men upon whom the families were relying for information, didn't even know of it. It was only when Whittall went to visit one of the two survivors, Russell Smith, on Sunday that he learned of the footage. Catriona Bayliss, who was with him, noted in her diary: 'PW was disappointed he did not know about [the CCTV footage] when everyone else seemed to know about it.'[27]

Whittall asked for a copy to be delivered to him in Greymouth. A memory stick with the footage arrived at his base at the Coleraine Motel on Monday.

The prime minister returned to Greymouth on Monday morning. It was now day three – November 22 – and the government had become aware that some of the families were directing their rage at the police and accusing them of blocking the Mines Rescue men from going in. Key's mining knowledge had grown a little since the Friday evening of the explosion. He now understood that there were few similarities between the dramatic rescue of the 33 Chilean gold miners and the entrapment of 29 men in the volatile and gaseous maw of the Pike River coal mine.

Key asked to be taken to the mine site, where he briefly addressed the brigadesmen assembled in the Mines Rescue room. He then said, 'I'm going to come around and talk to each of you, and if any one of you think we should be going in there you need to tell me, because whatever you tell me I'm going to tell those family members.' He circled the room, shaking each man's hand. All of them said: 'Look, don't go yet. It's not right to go yet.'

A couple of the men asked if he had seen the CCTV footage of the explosion. It was the first he'd heard of it and he asked to be shown it. Whittall was in the control room with Key as it was played. 'I just went "Hell!" because it was pretty obvious from the video that the explosion was massive,' Key would recall. Shocked, he asked for the footage to be played again a number of times. He couldn't see how anyone could have survived the blast.

The company men, though, continued to maintain that the footage didn't eliminate the chance that men could still be alive. Key told them they had to show the footage to the families and the media. 'I said, "I'm the prime minister – I can't un-know things." There was quite a bit of debate.'

As they left the mine, Key and his entourage met an incoming bus carrying family members to the site. Asking his driver to stop the car, he boarded the bus and spent several minutes speaking to them. His message was direct: 'I went [to the scene] to ask the Mines Rescue Trust, as individual men, whether they were prepared to go in the mine, and whether there was any daylight between them and what the police are telling you. I have got to tell you there is none.'

Key didn't mention the CCTV footage, or his impression that the blast was so powerful it must surely have killed the men. He felt he had made it clear to the company that he expected them to make the footage public.

That night the president of Chile phoned to offer his country's assistance and equipment, should it be required.[28]

The families were not shown the CCTV footage at Monday afternoon's meeting, although it was shown to a briefing of officials at three o'clock that day. Nor were they shown it at Tuesday morning's meeting. Knowles felt care needed to be taken how it was shown, and that its implications had to be properly explained. He was conscious that almost half the men underground were contractors, whose anxious relatives were not knowledgeable about mining.

There was, however, a subtle shift in the tone of the official discourse. The *Greymouth Star* reported on Monday afternoon that Whittall hoped the men would demand to know of their rescuers, 'What the hell kept you?' However, 'the reality is it's been three days. It's becoming more and more grave with every hour for the families'.[29]

From Knowles there was an admission that there could be loss of life, but he remained optimistic.[30] In reality, he felt there was no hope of survivors by that time, but Whittall's tone of reassurance was making it impossible for anyone with accurate information – including families with contacts at the mine site – to stand up and pierce the hopes of those who still believed their men would emerge alive.

On Monday night there was anticipation among the families: an army robot mounted with a camera was to be deployed into the tunnel the next morning. It was hoped it would deliver the first images of conditions underground. But Tuesday morning dawned with bad news: 550 metres in, the robot had got wet and broken down. Tony Kokshoorn bellowed to the media that someone ought to have thought to put some plastic wrap over it. Later that night a second robot managed to travel just over a thousand metres up the tunnel and take photos of Russell Smith's abandoned cap lamp, which was still shining. It was 'good news for the guys underground' – they would still have light if they had used their lamps sparingly, Whittall said.[31]

By Tuesday morning, journalists had learned of the existence of CCTV footage. Some Pike employees who had been called back to help at the mine that morning had been shown the film; word was passed to *The New Zealand Herald*, and deft questioning of John Key flushed out confirmation that it existed. A few hours later, at the afternoon meeting – four days after the explosion – the footage was finally shown to the families and then to the media. When Whittall was asked why it hadn't been shown earlier, he said the 'piece of information just hasn't been relevant to doing a rescue operation.'[32]

Many who saw the footage were stunned by the violence of the explosion. Whittall acknowledged at the media conference that there could be deaths. 'I think it is becoming obvious there may not be 29 guys all sitting together waiting to be rescued. How many of them are I don't know, but those are the ones I hope are rescued and those are the ones I am waiting to see.'[33] Gary Knowles described the outlook as 'bleak' and 'grave'.[34]

Behind the scenes, the relationship between Whittall and Knowles had become severely strained. Knowles believed that Whittall was undermining his position. Before each meeting with the families the two of them would agree on the content and tone of the briefing, but Knowles felt Whittall would then deviate from their agreement and inflate the family's hopes.

In turn, Whittall was making his displeasure with Knowles' leadership known in high places. On the Tuesday morning he complained to Gerry Brownlee, saying he was having 'ongoing difficulty' trying to contact and brief Knowles.[35] By then, Knowles had told Whittall that a police investigation into the explosion was underway, and that the contents of computers at the mine site would be cloned.[36]

John Key was privately questioning the 'heroic assumptions' that conditions underground could be survivable.[37] Quiet discussion was underway among police and officials about when to stop calling the operation a 'rescue' and start calling it a 'recovery' effort.

Troy Stewart, one of the first Mines Rescue men on the scene on the Friday night, had, like his colleagues, spent every waking hour since focused on the operation; he worked at the mine through the night, went home in the morning to try and get a couple of hours sleep, and then headed to the rescue station at Rapahoe until it was time to return to the site for his next shift. There was no time to read the papers, listen to the news, or talk to friends about what was going on.

Only Stewart's wife Jolene saw how angry and frustrated he was with the conduct of the operation. He had known ever since he saw the CCTV footage on the Friday night that the men were dead, and he believed that preparations to seal the mine should have started quickly, before it exploded again. But up at the site all sense had been destroyed by those who clung to the notion that survivors were waiting. It had been more than four days; not only was it almost certain that a fire was burning, it had also been established on the Monday that the compressed air line – the lifeline the families were being told men could be sucking on – had ruptured somewhere between 1.6 kilometres and two kilometres into the stone tunnel, and would therefore not be reaching into the mine workings.

'No one was listening,' Stewart recalls. 'The police wouldn't listen. The mine staff wouldn't listen. The police were being swapped out and new ones would come in, and they'd have to be re-educated. ... It was a certainty that it would blow again. We knew it had to be sealed but we weren't allowed to do it.'

On Wednesday morning he came home from his overnight shift and happened to pick up a newspaper from the previous day. He erupted in rage: it was the first time he fully understood what a false picture the outside world was getting. He tore it up, flung the pieces across the room, and went out on his bike up the coast road towards Westport to try and calm his mind.[38]

While Stewart was out of range, word began circulating in town that conditions in the mine had stabilised and there was to be a re-entry attempt by Mines Rescue. Jolene Stewart heard the news while she was at work at the Greymouth Paper Plus bookstore.

Shortly after midday Whittall was phoned by Knowles from the mine site and told that Mines Rescue men were preparing to go underground.[39] He flew up by helicopter, along with Bayliss and security man Nick Thompson.

In fact there was no imminent rescue attempt, but there was a swollen sense of expectation. Early that morning the first new borehole had punctured into the heart of the mine. The progress of the drill bit had been monitored by the nation through news reports over the previous two and a half days, and everyone at the scene was clinging to hope that it would provide the information needed to enable a re-entry effort. News that the first samples drawn were virtually fresh air produced a euphoric reaction, although Trevor Watts and Darren Brady cautioned that at least three hours of readings were needed to create a representative picture of the conditions.

Early in the afternoon two mines rescue teams were briefed by the Police Disaster Victim Identification experts on what to expect when they encountered the bodies and remains of deceased miners. At the same time, experts with laser scanning equipment capable of capturing three-dimensional images from underground cavities were up on the hill, having toiled since mid morning to gather pictures from the bottom of the slimline shaft. Gas sampling from the new borehole was continuing, and results had been emailed to gas expert Dave Cliff at the University of Queensland for review.

At 1.17 Knowles advised Grant Nicholls that gas readings suggested there could be a window of opportunity to enter the mine at three o'clock; a risk assessment would be available at two.[40] The process of getting risk assessments signed off by Wellington had become the subject of angry frustration at the site. Knowles was given the clear message that, if conditions were right for a re-entry attempt, there must be no undue delay in getting the necessary approval from the senior bureaucrats.

Mines Rescue Service general manager Trevor Watts was meanwhile meeting with Steve Ellis and two Australian experts: Ken Singer of the Queensland mines inspectorate and Seamus Devlin, head of the New South Wales Mines Rescue Service. The group were working on a document setting out the controls that would need to be in place before brigadesmen could be sent into the mine. If there were to be a re-entry attempt, Watts thought it was unlikely to occur until the early hours of the following morning as the mine atmosphere was more stable at night. The pattern of readings over the previous few days showed that at between two o'clock and four o'clock each afternoon methane levels usually reached explosive levels.[41]

While they were meeting, an email came through from Dave Cliff in Brisbane. His analysis of results from the new borehole had showed there was a methane fire burning underground. Cliff had heard that a rescue attempt was being considered and he was desperate to make sure that no one was about to enter the mine.[42]

Cliff's results were conveyed to an incident management meeting and there was heated debate about the implications. Whittall had arrived at the site by helicopter at about two o'clock. While he waited for clarification on the gas results he went into the control room and asked Barry McIntosh to show him again the footage of the Friday explosion.

Up on the hill, meanwhile, the laser scanning crew led by John Taylor, a world expert in the technique, had completed their task of lowering a camera down the slimline. They had packed up and were waiting near the shaft to be picked up by chopper. Suddenly they heard a loud roar and took off downhill. A huge plume of smoke, soot, coal dust and other debris had erupted up the slimline shaft. As it dissipated they could see an enormous pall of smoke hanging over the top of the main ventilation shaft. Debris had crashed down and smashed some of their equipment. They were covered in dust and soot, but miraculously unharmed.[43]

Whittall was watching the historic footage of the Friday explosion when word came through to the control room: a helicopter pilot had seen an explosion erupt from the ventilation shaft. Barry McIntosh immediately switched over to the real-time signal from the CCTV camera at the portal. Knowles, told of the explosion, rushed to the control room. He and Barbara Dunn, along with Whittall, Thompson and Ellis, watched the second explosion play out on the computer screen. There was a 30-second surge of air out the portal; the air then sucked back in, before blowing out again. It was even more violent than the explosion of Friday the 19th.

McIntosh turned to the police officer and saw tears well in his eyes.

The group moved to another room, where Ellis instructed them on the implications of the explosion. Referring to a mining handbook he had with him, he explained the mechanics of the blast and how it would have lost velocity with each 100 metres that it passed down the long 2.3-kilometre stone tunnel to the portal. The message was clear: the blast was completely unsurvivable.

In the room where the Mines Rescue men had based themselves for five days, decked out in their rescue clothing and constantly checking and rechecking their breathing apparatus and equipment, Trevor Watts broke the news of the second explosion. Upwards of 24 brigadesmen were present. The men released a collective groan of defeat and grief, laced with profound relief that they had not been on the way up the gun barrel when the mine blew again.

Down the valley in Greymouth, deputy mayor and Paper Plus owner Doug Truman came into the shop with news of the second explosion. Jolene Stewart was desperate, fearing the brigadesmen had gone in and her husband had been among them.

Troy Stewart was returning to Greymouth from his long and angry bike ride, and called into the rescue station on his way home. His colleague, Glen Campbell told him as soon as he walked in the door: 'It's just gone again.'

Whittall wanted the families informed of the explosion before they got word through the media. He, Catriona Bayliss and Nick Thompson boarded their helicopter and headed back to Greymouth.

Knowles and Barbara Dunn travelled back by police car with Inspector Dave White. They discussed how the news should be conveyed to the families. Dunn sat in the back of the car and wrote a short script. It was clear there must be no false hopes raised after this second immense explosion: the news needed to be delivered clearly and without ambiguity. She wrote out two copies, one for Knowles and the other for Whittall.

A family meeting had already been scheduled for 4.30. At 2.55 a text was sent to all those on the communications list. It read: 'There will be a significant update at the 4.30 family meeting. It is recommended that all families attend.'

About 500 people assembled, including young children. Having heard the buzz about town that a rescue attempt was in the offing, many thought the text message carried a hint of good news. Others arrived with dread in their hearts. A large number of uniformed police officers lined the large noisy basketball court.

Peter Whittall, Gary Knowles, Barbara Dunn and Gerry Brownlee met in the carpark before entering the Civic Centre. Knowles and Dunn felt it was preferable for Whittall to deliver the news. The families seemed to trust him, while Knowles had increasingly been the target of venom from anguished family members. However, Whittall was told that if he felt unable to do it, Knowles would take over.

Whittall felt he should be the one to tell them. He took the piece of paper on which Dunn had written the short and tragic script. The group entered the building.

Whittall began to speak. As he was inclined to do, rather than launching immediately to the point he began by explaining the context. He said that gas levels earlier in the day had shown some improvement, and he had been called to the mine because the Mines Rescue men were preparing to go in.

People instantly began to cheer and clap.

Knowles, Brownlee and Whittall waved their arms for silence, realising the message was going terribly wrong. Whittall tried to go on but was unable to. The assembled family members quickly picked up on the uncertainty; pandemonium began to break out. Knowles stepped forward and broke the news that there had been a second explosion and no one could have survived.[44]

The crowded hall erupted in a cacophony of grief. Some people wailed; some fell to the floor. Others shouted, screamed and swore. Some ran from the room. Others verbally abused Knowles. One man tried to throw a chair at him. Whittall, plainly distressed, hugged some of the family members while Knowles was quickly ushered out of the room.

It was over, but there was no end in sight.

For five days the families and the nation had been encouraged to believe there was genuine hope when there was none. For five days, a cast of many – Pike employees, police, firemen, Mines Rescue men, visiting experts – had been privy to information that disclosed the likely death of the men, but the families of the 29 had not been told.

Now in one terrible clumsy moment, all hope was extinguished. Family members, adrift on an ocean of grief, staggered from the hall and into the glare of the international media.

And yet there was still only the idea of death. There was no evidence to see: no visible accident site, no bodies of the beloved, no visceral proof. There was just the solidarity of a nation in shock and the fierce embrace of a community that mustered its strength to hold its people up.

TWELVE
Entombed

For more than a decade, ever since the mine project was initiated on the ill-founded assumptions of its promoters at New Zealand Oil & Gas, Pike River Coal had made grand promises and failed to deliver on them. In the five years the mine had been under development, it had lurched from one miscalculation and mishap to the next, while continuing to boast of handsome returns to come. In the few short months that it had managed to produce a little coal for export, it had posed as an exemplar of modern health and safety management, when in reality it was skating at the edge of catastrophe.

Even now, in the teeth of utter tragedy and failure, the pattern of promise and betrayal continued.

A day after the second explosion, Peter Whittall vowed to the families that the bodies of the men would be brought out of the mine. 'We've given an undertaking … we'd give them their family back again.'[1] Neither he nor Pike River Coal was in a position to make such a pledge: the company was fast going broke.

In the weeks leading up to the explosion, Pike's wages and bills had been funded by New Zealand Oil & Gas under the $25 million short-term loan facility negotiated in late September 2010. The mine was on the brink of being refinanced yet again: on November 19 Pike director

Stuart Nattrass had concluded a deal that would see finance house UBS underwrite a $70 million equity raising.

The week of November 22 UBS was to talk to investors to confirm their interest in subscribing for shares. With the international price of hard coking coal surging again, there was little doubt the offering would receive solid backing. The injection of fresh capital would enable Pike to pay off its debt to NZOG, and leave enough in the bank to cover the mine's weekly costs until the end of February 2011, after which it was expected that receipts from coal sales would be flowing in.[2]

Now the disaster meant the $70 million capital-raising was off, and there was no guarantee that NZOG would continue providing cash under its short-term lending facility. By the time of the first explosion, Pike had drawn down $13 million of the $25 million available under the NZOG loan. But the deal contained a force majeure clause that freed NZOG of any legal duty to make the remaining $12 million available.

The directors of Pike River Coal were in a difficult position. If they allowed the creditors to be paid and continued to ask contractors to undertake work in connection with the emergency, they could be liable of trading the company while it was insolvent.

The payments – which ought to have been received into the bank accounts of the builders, pipe-layers, mechanics and sundry other contractors the week of November 22 – were withheld. Even firms whose men were dead in the mine, such as Milton Osborne's SubTech, were not paid.

Almost a week after the first explosion, New Zealand Oil & Gas advanced the balance of the $25 million short-term loan, stating that it was the 'right thing to do commercially and ethically'. Pike could now at least pay its ongoing wage costs – but there was still nothing for the small local contractors and other unsecured creditors who had invoiced for services prior to the explosion.

Following the second explosion on Wednesday, November 24, discussion proceeded about how to bring the mine atmosphere under control so the bodies could be recovered. There was general agreement that the best option was to use a Górniczy Agregat Gaśniczy (GAG) machine, a device that pumps inert gases and water vapour into mines to extinguish fires and stabilise the atmosphere.

The Queensland Mines Rescue Service had placed its GAG machine at the ready on the night of the first explosion, and waited for a request to send it to New Zealand. By November 23 – four days later – Doug White had concluded the men were probably all dead, and at a meeting at the Greymouth Police Station he urged that the GAG be deployed from Queensland. He predicted that without prompt action to make the mine atmosphere inert it would explode again, probably repeatedly.

Preparations were begun to bring the GAG machine to New Zealand, but White was told by police they 'didn't want it in the car park' as this would send a message to the public and the families that there was no longer any hope.[3]

The machine arrived in New Zealand on an Air Force Hercules and was transported up to the mine on Friday, November 26. Various matters had to be attended to before it could be put to work. There were risk assessments to be approved by Wellington. There was paperwork to sign off declaring that all life inside the mine was extinct and that therefore it was appropriate to take steps to make the mine inert. And enough aviation gas had to be sourced to satisfy the machine's enormous thirst.

Trevor Watts and Dave Stewart of the Mines Rescue Service were meanwhile struggling to impress upon the assembled police and Department of Labour officials that the situation was urgent. If the atmosphere were not brought under control rapidly, the mine would blow up again and again.

They were proved correct. On the Friday that the GAG arrived at the site the mine exploded for a third time, almost exactly one week after the first explosion. On Sunday, November 28, it exploded again. Flames fuelled by Pike's thick seam of premium coking coal leapt from the top of the 111-metre ventilation shaft. The entire surface fan structure was blown to one side and demolished. It looked as if the whole forested wilderness of the Paparoa Range might be set alight.

With fire continuing to erupt out of the shaft, the families asked if the bodies of the men would be home by Christmas. Peter Whittall replied that it could be 'weeks' before that happened.[4] John Dow, at the helm of a company whose single asset was on fire with 29 bodies inside and which couldn't pay its bills, commented to the *Greymouth Star* that it was 'definitely feasible' to get the mine working again.[5]

The Mines Rescue men were beside themselves with frustration. Hemmed in by the lumbering processes of the police and the Department of Labour, Trevor Watts warned that the chance to recover the men's bodies may have been lost: there was evidence that the fourth explosion had caused a roof collapse somewhere along the main tunnel and this would block the passage of brigadesmen. Yet the families' hopes that the bodies might soon be returned to them were still being raised.[6]

When a national memorial service was held under a brilliant blue sky at Greymouth's Omoto Racecourse on December 2, the Mines Rescue men travelled together by bus and stood as a team. But there was no sense of finality. Many were angry and bitter they hadn't been able to take the steps they believed were necessary to control the mine atmosphere when the evidence of a fire underground was first presented by Robin Hughes, less than 24 hours after the first explosion. And myriad complex and risky tasks lay ahead: they were helping with the GAG machine, assisting with the placement of containers and concrete across the portal to stop the flow of oxygen into the mine, and working on plans to lower two gigantic metal semicircles by helicopter over the top of the ventilation shaft.

Each step was hopefully taking them closer to stabilising the mine so they could gain entry and search for the bodies.

At the service, dignitaries, politicians and Pike River executives occupied the dais while the families were seated in the racecourse grandstand. A crowd of 10,000 sat under the hot sun. With the gentle shape of the Paparoas in the distance, a Greymouth singer and songwriter, Paul McBride, sang 'Brothers 29', the soft, sad folk melody he had written for the lost men. Pipe and brass bands from Greymouth, Hokitika and Westport played. The children of Greymouth's combined primary school choir sang 'Mawhera', a hymn to the West Coast. Local writer Helen Wilson read 'We Will Live', a poem she had written for her shattered community. Greymouth singer Carolyn Williams and her daughter Sarah sang 'You'll Never Walk Alone'. The choir of St Patrick's Church, where Peter and Leanne Whittall had worshipped, sang 'Amazing Grace'.

The names of the 29 dead were read aloud: the list went on and on. Across the country, New Zealanders watched the service live on their television sets in the middle of a Thursday afternoon, and wept.

Over the weeks, sympathy and support flowed into Greymouth. Air New Zealand deployed its team of specially trained crisis support workers and assigned one member to each distraught family for the first week after the explosion. Mayor Tony Kokshoorn set up a fund for the miners' families, and $8.2 million flooded in from around the nation and the world. Companies and individuals sent donations ranging from butter and pork to petrol vouchers and boxes of children's toys. In February 2011 the Warriors and the Newcastle Knights arrived for an international class rugby league match. They played at Wingham Park, the Coal Creek ground on which Pike miner Blair Sims had scored so many tries in his stellar sporting career; a crowd of 5,000 came together in a day of communion and mourning.

On December 13, less than three weeks after the first explosion, Pike's failure was complete. The company was placed in the hands of receivers John Fisk, Malcolm Hollis and David Bridgman of Pricewaterhouse-Coopers. Any shred of dignity the company and its mine project still had was now gone. On the books, the Pike development was worth $340 million, and 180 people had been employed in what was to have heralded the beginning of a new age of prosperity for the West Coast. Now, with 29 dead and the asset out of control, the creditors needed their money back. New Zealand Oil & Gas – the company that had promoted the mine and funded its development on the sweat equity of consultants, the coal-hungry capital of Gujarat NRE and Saurashtra Fuels, and investors who saw Pike as a chance to participate in the international resources boom – was owed $64 million in loans and had $82 million tied up in Pike's now worthless shares. The Bank of New Zealand was owed $23.5 million.

The receivers immediately sacked 114 Pike workers. Among those who were retained to help manage the asset were chief executive Peter Whittall, statutory mine manager Doug White and production manager Steve Ellis. Ellis, who had failed three times in Australia to obtain a first-class mine manager's certificate of competency[7], became qualified to hold the position of mine manager in New Zealand a month after the disaster. In May 2011 he would replace Doug White as statutory mine manager, in the employ of the receivers.

Dozens of local businesses, including those who had lost employees in the mine, were unsecured creditors and therefore at the back of the queue for repayment; they ranked behind the secured debt owed to New Zealand Oil & Gas and the Bank of New Zealand. There appeared to be little prospect that they would be paid the millions of dollars they were collectively owed by Pike River Coal.

The money included the wages of workers employed by contracting companies who had provided underground services to Pike. Shocked and grief-stricken, the contractors now had to combine their resources and do battle with the receivers. Under the leadership of Grey District councillor and local businessman Peter Haddock, they argued that the receivers had a moral duty to at least pay the five contracting firms who had lost men in the mine.

The receivers went part of the way: they agreed to pay the wages due to the dead employees of the five firms, as well as to their surviving colleagues who had worked in the mine.

Insolvency laws were against the contracting firms. A campaign to publicly shame the receivers and New Zealand Oil & Gas and the Bank of New Zealand was the only strategy likely to work. With the help of prominent Wellington lawyers David Butler and Greg King and former West Coast journalist Gerry Morris, pressure was applied through the media and lobbying directed at the prime minister's office.

The campaign was partially successful. New Zealand Oil & Gas eventually agreed to forego some of the $80 million paid out under Pike River Coal's main insurance policy. This enabled payment of the first $10,000 owed to each unsecured creditor, and 20 cents out of every dollar owed beyond that. It took the contractors' group almost 12 months to secure the deal. And many small local businesses were still owed hundreds of thousands of dollars.

The families of the Pike 29 also quickly grasped the need to organise themselves. Soon after the second explosion Bernie Monk, whose son Michael had been killed in the mine, phoned his brother-in-law, Greymouth lawyer Colin Smith. 'There's something very wrong here,' he said. 'We need to get a team together.'

Within a few days, all 29 families were represented on the group. Monk, a gravel-voiced publican whose family owns the Paroa Hotel just south of Greymouth, became the spokesperson. Carol Rose, who lost her son Stu Mudge in the mine, became the secretary, and Smith became chair. The group met every week, at first in the expectation that the bodies of their men would be returned to them, and then in the growing awareness that every step of the long journey ahead would be a struggle. They would have to fight to ensure the mine was not simply sealed over and the bodies of their men abandoned. They would have to fight for information and answers. And they would have to fight for remorse and accountability from those who had allowed the mine to march, unhindered, to calamity.

The first brutal lesson came on January 13, 2011, less than two months after the first explosion. New Zealand Police had been in control of the emergency and recovery operation from the beginning and was paying the huge cost of running the GAG engine, drilling boreholes, monitoring gas levels, flying helicopters on missions to the ventilation shaft, and pumping nitrogen into the mine.

The GAG had been running night and day since December 1 – the duration of its operation at Pike was beyond precedent – and the machine had been almost thrashed to pieces. It was at risk of flying apart in a catastrophic failure. The Queensland Mines Rescue Service wanted its machine back by the end of January with the wear and tear made good; to this end the police had to source and foot the bill for two replacement engines.[8] A Floxal machine had also been working since December 18 in an attempt to flood the mine with nitrogen.

Despite all this, no one had yet been able to get underground, and the police had reached the conclusion it was probably impossible. The commissioner, Howard Broad, told receiver John Fisk he had lost confidence that anyone would safely re-enter the mine. In any event, the police had been advised by the Crown pathologist that the prospect of finding recoverable human remains was low.[9]

An $11 million plan for a staged re-entry into the mine, put together by Doug White just before Christmas 2010, had been comprehensively shot down by the man hired by the police to review it. David Reece, an Australian mine safety consultant, had pointed out that the very features that had made Pike such a flawed mine were a fundamental obstacle to getting back into it. There was only one way in and one way out – via the long stone tunnel, a design that 'would not be accepted in most developed underground coal mining nations'.

'To mount a staged re-entry, then control of the (as yet undetermined nature and location) fire or hot coal condition, with only this single 2.5 kilometre escape route, even if in good condition, is high risk,' Reece reported. 'The roadway is anticipated to be littered with twisted steel from the subsequent explosions, no simple task to negotiate under breathing apparatus. The GAG unit will have potentially caused significant damage to the strata, not easily recovered under mines rescue conditions. ... This mines rescue resource requirement will be substantial and of considerable duration.'[10]

In an internal police file note, Howard Broad recorded 'justifiable concerns' about Pike River Coal's 'overly optimistic views about achieving re-entry into the mine. On a number of occasions representatives of

PRC had demonstrated a lack of realistic understanding of the likelihood of being able to achieve a safe re-entry ... in the reasonably foreseeable future.' There were questions about Pike's 'appreciation of risks associated with this mine'.[11]

The Fire Service, too, was pessimistic. Jim Stuart-Black, its national manager of special operations, told others reviewing the plan that 'the risks associated with any re-entry to recover the deceased are too high and few (if any) other countries would allow such an operation to take place.'[12]

The police had other worries too. Broad noted that 'any action by police or the company with respect to the physical state of the mine may well have implications for the only other material asset of the company – its insurance policies. ... Police and the wider Crown must be careful to ensure they have authority for their actions on site and do not create liabilities to the company, its secured creditors or the insurers.'[13]

And there was the issue of liability under the Health and Safety in Employment Act now the police's emergency management role was over. Superintendent Gary Knowles wrote to Broad: 'Currently Police control the mine site and the operation. As a result of that control, Police [face] potential liability if anyone working on the operation is exposed to unreasonable danger.'[14]

On the morning of January 13, Bernie Monk started receiving calls from reporters who had heard the police were about to pull out of Pike. Monk and the other families knew nothing. Later that day Howard Broad, along with David Reece and Gerry Brownlee, flew to Greymouth to make a shock announcement: the mine was to be sealed and handed over to the company's receivers.

The announcement met with instant resistance. The family group had heard from contacts at the mine site that the GAG machine had succeeded in stabilising the atmosphere, and that footage from underground showed far less damage to the mine workings than expected.

Trust in Pike River Coal had already evaporated, and now it seemed that even the police were withholding information. This impression was later reinforced when Monk received a tip-off: three-dimensional images taken from the bottom of the slimline shaft shortly before the second explosion and handed over to the company and the police showed that a box containing self-rescuers had been opened.

The revelation raised vital questions. The chief coroner, Judge Neil MacLean, had ruled that all 29 men would have died either at time of the first explosion or within minutes, but perhaps this was wrong? Perhaps men had survived long enough to swap their spent self-rescuers for fresh ones? Perhaps they had survived the explosion but then been unable to get out?

The family group asked to see the images. A meeting was arranged at Hornby Police Station in Christchurch and Monk and other representatives viewed the footage. A few days later there was another stunning revelation. The police advised that footage taken down a borehole in February revealed a fully clothed and intact body lying in an area close to where other men were known to have been working at the time of the first explosion. This proved that, despite four powerful explosions, there were remains to be recovered, mourned and buried.

Although the families had managed to block the January plan to permanently seal the mine, in March 2011 the site was handed over to the receivers. These people had no interest in attempting to re-enter the mine and recover the men's remains. Their job was simply to get the best possible price for the critically damaged asset. There was talk of plans to stabilise the tunnel, but by May it was clear that nothing was being done. Supported by their lawyers Nick Davidson, Richard Raymond and Jessica Mills, the families were granted a high-level meeting with the receivers, police and government officials. They won a commitment that a feasibility plan would be developed for a staged re-entry into the mine.

In July the Mines Rescue Service succeeded in constructing a temporary seal 170 metres in from the portal, and by August the brigadesmen believed conditions in the stone tunnel were good enough to attempt a 'reconnaissance walk' under breathing apparatus as far as the major roof collapse at the end of the tunnel. Steve Ellis, then statutory mine manager in the employ of the receivers, blocked the move and came up with an alternative plan. He promised: 'We will reclaim that tunnel before Christmas, I'm quite confident of that.'[15] The tunnel was not 'reclaimed' by Christmas 2011, or even by Christmas 2012. By September 2013 no one had gone beyond the 170-metre seal. The families' deal had turned out to be yet another mirage.

In 2012 the receivers sold Pike to the government-owned coal mining company Solid Energy Ltd: the mine that was valued at $340 million on the books was handed over for $7.5 million.

For the families, the sale was the first positive news they had had for 18 months. 'Solid Energy has said in the past that recovery is their main priority and that is a giant step for us,' Bernie Monk told the media.[16]

Their relief was short-lived. The sale had been contingent on the new owner's commitment to take 'all reasonable steps' to recover the men's bodies as part of any commercial mining operation at Pike – provided it and the government considered recovery could be achieved safely and was 'technically feasible and financially credible'.

Solid Energy's chief executive, Don Elder, stressed there would be no stand-alone operation to retrieve the men's remains. Any recovery attempt would occur only as part of a renewed mining operation at Pike. And given that the original Pike operation had been based on skimpy geological data, and that the underground infrastructure had likely been destroyed in the explosions, the new owner would have to start from scratch – drilling enough boreholes to build a reliable picture of the geology, designing the underground infrastructure in a way that enabled coal to be safely extracted from the gassy seam, and bringing in new machinery. Elder put the odds of that happening at just five to ten percent.[17]

In the months after it signed the deal to buy Pike, Solid Energy itself descended into a crisis triggered by a sudden collapse in the international price of hard coking coal. Teetering on the edge of bankruptcy, the company mothballed its Spring Creek mine and laid off more than

600 miners, white-collar workers, and staff involved in alternative energy projects. In early 2013 an embattled Don Elder quit.

The Pike family group fought on, driven by the energy and anger of Bernie Monk, the calm resolve of Carol Rose, and the pro bono support of lawyers Smith, Davison, Raymond and Mills. It hired an engineer, Bruce McLean, to develop a plan to re-enter the mine and re-establish it as a commercial operation. The plan was rejected by Solid Energy. They gained the support of three experts – former United Kingdom chief mines inspector Bob Stevenson, British mining engineer David Creedy and New Zealand mine safety consultant Dave Feickert – who developed a proposal for bypassing the rockfall and gaining access to the inner workings of the mine where the bodies lay. The proposal was not taken up. They continued to keep open the possibility that one day something of their men would be returned.

Through the winter and spring of 2011 and early summer of 2012, the Royal Commission on the Pike River Coal Mine Tragedy sat at the Greymouth District Court. Verbal evidence was given by 51 people: experts of various stripes; former Pike personnel, including Whittall, White and Ellis; former chair John Dow; miners; Mines Rescue men; bereaved family members.

Nothing was heard from other people who had been closely involved in the development of the mine. Gordon Ward, who had been Pike River Coal's chief executive until two months before the disaster, remained resolutely in Australia, where he now lived. Tony Radford, who had chaired the Pike board until 2006, remained a director until June 2011, and was chair of New Zealand Oil & Gas, wrote a brief submission under a compulsion order from the Royal Commission. He was not called to give testimony.

Almost 70 lawyers were hired by parties with an interest in the proceedings. The cost of the commission's work came to $10.5 million.

Every weekend during the hearings, Nan Dixon set to work in her Rūnanga kitchen. In a frenzy of baking, she whipped up enough slices, cakes and biscuits to sustain the family representatives working on behalf of the Pike 29; it was something constructive she could do to help.

Every day through the long hours of testimony, the same worn faces would be seen in the public gallery, determined to make sense of the loss of their husband, son, father, brother, uncle, friend. And every day they would hear dark new evidence revealing the immensity of Pike River Coal's failings. Mining a coal seam known to be high in methane gas, the company had not installed a system fit for the purpose of monitoring its major hazard. Fixed sensors in the mine hadn't been working for weeks before the explosion: one had been out of action since September 4; another since October 13; one was unreliable and transmitted incorrect gas readings after it had been 'poisoned' by high methane levels, causing a flat line to show on the surface control room system. Nothing was done, despite statutory mine manager Doug White having signed off a report stating the monitor was 'stuck' at a reading of 2.8 percent methane.

They heard how Pike's electrical system departed from conventional mine design: the main fan – the principal source of clean air for the workers – was placed underground, where it could not be reached in the event of a disaster. The fan's motor and other items of electrical equipment underground were not designed to be intrinsically safe or flameproof, and could operate only in fresh air.

Despite early plans to pre-drain methane from the coal seam before mining began, this hadn't happened. Instead, gas was bled out of the seam from the in-seam drill holes that Pike relied on to find out where the coal seam lay, and drained through a pipeline the company knew to be inadequate and overpressured.

Pike had forged ahead with the introduction of its hydro-mining system before establishing the second means of egress that was required in law. It had widened the area to be mined with the hydro monitor without adequate knowledge of how the strata above would behave. It had started commissioning the system without having sufficient skilled workers to man it, and in the face of repeated spikes in the volume of methane released into the mine atmosphere.

The evidence laid before the Royal Commission showed that day after day Pike's underground workers – miners and the many contractors who had no background and little training in coal mining – had walked into an environment that might have exploded any number of times.

'There were numerous warnings of a potential catastrophe at Pike River,' the commission said in its final report. One source of this information was the reports made by the underground deputies and workers. For months these men had reported incidents of excess methane, and many other health and safety problems. 'In the 48 days before the explosion there had been 21 reports of methane levels reaching explosive volumes, and 27 reports of lesser, but potentially dangerous, volumes of the gas,' the commission noted. 'The reports of excess methane continued up to the very morning of the tragedy. The warnings were not heeded.'

Pike River Coal was focused on its short-term needs, the commission concluded. The mine was badly behind schedule and its credibility was under strain; it had faced difficulties from the start of the project, promised production volumes had proved illusory, and it needed more capital. There was a drive for coal production before the mine was ready. By November 2010, 'the emphasis placed on short-term coal production so seriously weakened Pike's safety culture that the risks of an explosion either went unnoticed or were not heeded.'[18]

Those whose duty it was to manage the risks of the operation had failed. 'In the drive towards coal production, the directors and executive managers paid insufficient attention to health and safety and exposed the company's workers to unacceptable risks. Mining should have stopped until the risks could be properly managed.'[19]

And the regulator had allowed Pike to continue operating in breach of the law. 'The Department of Labour did not have the focus, capacity or strategies to ensure Pike was meeting its legal responsibilities under health and safety laws,' the commission found. 'The department assumed Pike was complying with the law, even though there was ample evidence to the contrary. The department should have prohibited Pike from operating the mine until its health and safety systems were adequate.'[20]

But the commission – comprising Justice Graham Panckhurst, a pragmatic and sensitive High Court judge who had grown up in the West Coast town of Reefton, Stewart Bell, Queensland's commissioner for mine safety and health, and David Henry, a former senior public servant – could not determine the cause of the explosion: the forensic evidence remained locked away in the inhospitable bowels of the Paparoa mountains. Assisted by the findings of a team of investigators hired by the Department of Labour and police and led by Australian mine safety expert David Reece, it could only come up with possible scenarios.

An enormous volume of methane had exploded, perhaps 2,000 cubic metres. The area most likely to contain this amount of gas was the goaf formed during mining of the hydro panel. The area was also intersected by an in-seam borehole, and this would have added to the volume of methane bleeding from the seam. The commission concluded that the most likely scenario was a roof collapse in the goaf that had suddenly expelled a huge amount of methane into the mine roadways and knocked over a nearby temporary stopping, with the gas becoming diluted to within the explosive range of five to 15 percent.

Another possible scenario was that a large volume of methane had accumulated in the area being worked by the ABM20 continuous miner and in other areas in the western part of the mine, perhaps because of a localised ventilation failure, or the intersection of an in-seam borehole. This theory was thought to be much less likely: mine officials in the area ought to have detected any increase in methane on their hand-held detectors, and readings taken in the hours before the explosion had not revealed elevated gas levels.

As to the ignition source that had set the methane aflame, there were several possibilities. A single spark was all that was required. Given that a pump had been switched on at about the same time as the explosion, the cause may have been electrical. Other possibilities were overheating of a diesel engine, or the use of illegal contraband such as a cigarette lighter, or a spark from the newly commissioned main fan.[21]

In the absence of hard evidence, the scenarios presented by the Royal Commission were just that. But whatever the cause of the explosion, one fact was obvious: it would not have occurred in an environment where the critical risks were properly controlled. 'Ultimately, all explosions are a manifestation of the failure of an organisation's health and safety management system,' the commission noted.

The Royal Commission published its findings in October 2012, almost two years after the first explosion. They brought relief but not resolution. The National-led government's minister of labour, Kate Wilkinson, fell on her sword in acknowledgement of her department's failure to bring Pike River Coal to heel for its multiple failings. But the company's directors and former managers publicly rejected the commission's criticism of their performance.

No one representing the company and its mine said sorry for their part in a calamity that had claimed 29 lives, scarred families and tormented a community.

The cruel tragedy of the event descended into a farce. The Department of Labour mounted an exhaustive investigation and laid charges under the Health and Safety in Employment Act against three parties. Peter Whittall was charged on 12 counts of acquiescing or participating in the failures of Pike River Coal, and of failing to take all practicable steps as an employee of the company to ensure that nothing he did (or didn't do) caused other workers harm. He pleaded not guilty to all charges.

Valley Longwall, the company hired by Pike to undertake in-seam drilling in the mine, pleaded guilty to three charges related to the operation of its drill rig despite Pike having failed for months to carry out essential safety checks. It was fined $46,800.

Pike River Coal was charged with nine offences under the Act. The failed company – still under the control of the receivers – did not put in an appearance at the Greymouth Court hearing, and did not present any evidence in defence or mitigation. Judge Jane Farish ruled against it in a decision that recorded 'an accumulation of errors and omissions which transpired over a number of years'. She found 'a systematic failure of the company to implement and audit its own (inadequate) safety plans and procedures'.[22]

When the judge came to impose sentence on the company, she sat at the bench struggling to control her tears. She had heard the statements of families still locked in despair more than two and a half years after the disaster. Anna Osborne spoke of her depression and inability to move on until the remains of her husband Milton were recovered. Willie Joynson's widow Kim told of the suffering and ill-health of her children following their father's death. Sam Mackie's mother Beth described the horror of losing her only child. Kath Monk spoke of the pain at being robbed of her son Michael, and her disgust that no one from Pike River Coal had ever come to her and apologised for his death.

Judge Farish described the Pike disaster as 'the health and safety event of this generation. ... a worse case is hard to imagine.' But the futility of her task was palpable. The company she was about to punish had long since gone broke. The money gathered from the sale of the mine and a small stockpile of Pike coal, and from an $80 million insurance payout, had been distributed to creditors. There was only enough left in the pot to pay each family $5,000. She nevertheless ordered $3.41 million in reparation – $110,000 to each of the victims' families and to Daniel Rockhouse and Russell Smith. She also imposed a fine of $760,000 on the company for its multiple breaches of the law.

The company had demonstrated a total lack of remorse. 'It is not often a company steps back and holds its hands up and says, "I have nothing." Even a company in a fragile state usually comes forward and offers reparation, but here nothing has been forthcoming.'

She turned her glare on Pike's directors and major shareholder and said they could pay. But even the company's liability insurance – which, at $2 million, would have gone more than halfway towards the reparation order – had already been mostly spent on legal fees.

New Zealand Oil & Gas Ltd, the founding force behind Pike, responded to the judge's challenge by saying it had already gone above and beyond its legal duty: it had lent Pike the additional $12 million a week after the explosion when it could have turned off the cash tap; it had surrendered almost $7 million to enable the partial repayment of contractors who would otherwise have got nothing; and it had funded the receivership to the tune of $6 million, a process that had assisted in the 'orderly wind-up' of the company's affairs, when it could simply have liquidated Pike and taken back all that it was owed.[23]

On November 20, 2010, the day after the first explosion, Detective Superintendent Peter Read had been appointed to lead a police inquiry into the disaster. On July 17, 2013, he arranged to meet with the Pike families in Greymouth to tell them the outcome of more than two and a half years of investigation.

A team of up to 16 people, working in partnership with Department of Labour investigators and sharing an expert group led by David Reece, had conducted 284 witness interviews and reviewed the entire contents of Pike River's corporate hard drive. The key question: Was there sufficient evidence to charge any individual with the manslaughter of the 29 men?

Read's team had assessed the potential culpability of several individuals who had held key management roles at Pike. They had also looked closely at an act of 'gross stupidity' committed the day before the explosion – the covering of a gas sensor on the Valley Longwall in-seam drill rig during an attempt to recover a drill rod stuck in the coal seam – but concluded that only a small volume of methane would have been released during the period the sensor was disabled.

There was ample evidence of gross negligence in the operation of the mine – methane sensors that didn't work, dozens of methane spikes that had not been investigated, insufficient ventilation to provide clean air to the number of places in the mine where work was going on, an inadequate system to drain methane from the coal seam, temporary stoppings that should have been upgraded to permanent structures but were not, flawed mine design. But the problem for Read and his team was that, because it was still not known what had caused the explosion, it was not possible to argue beyond reasonable doubt that the act or omission of any individual had led directly to the disaster. There were only well-educated guesses: the possibility of a roof fall in the hydro panel that expelled a huge volume of methane into the mine; a build-up of gas in the western part of the mine, which was prone to methane layering in the roof (including the ABM20 place, where the machine had intersected an in-seam borehole the morning of the explosion); the possibility of spontaneous combustion in the rider seam, the thin layer of high ash coal above the main Brunner seam.

The latter scenario had emerged late in the police investigation, after the Royal Commission had published its report. In early 2013 Read's team asked Darren Brady to bring a fresh pair of expert eyes to the police investigation. Brady, a gas chemist from Queensland's Safety in Mines Testing and Research Station, had helped at the scene following the explosion and had provided technical advice to the commission.

Brady had based the additional theory for the cause of the explosion – spontaneous combustion in the rider seam – on data obtained from the mine in the days after the explosions. Temperature readings taken at different points down the slimline shaft had shown it was much hotter around the rider seam than further into the mine. This had led him to wonder whether there may have been ongoing spontaneous combustion.

When the slimline shaft was drilled to help the mine's ventilation after the collapse of the main shaft in February 2009, Pike had inserted a metal sleeve to line the 600-millimetre hole. Brady thought it possible that the gap between the outside of the sleeve and the surrounding ground may have provided a pathway for oxygen to reach the rider seam, and that, over time, the oxygen had reacted with the coal to create spontaneous combustion. The combustion may have provided the source of ignition for the explosion, or it may have weakened the strata and caused a roof collapse that released a sudden surge of methane into the mine.[24]

It was just a theory, along with all the other theories. Without access to the underground scene, no one could be sure. None of the possible explanations could be ruled in or out, which meant that a prosecutor would not be able to prove beyond reasonable doubt that the negligence of any individual had caused the explosion and the deaths of the 29 men.[25] There was enough evidence to support a charge of criminal nuisance, but Pike's former chief executive Peter Whittall was already facing charges under the Health and Safety in Employment Act; there was a risk that any criminal nuisance charge would cut across Whittall's pending trial and create issues of 'double jeopardy'.

Read therefore advised the families that no one would be held criminally liable for the death of their men – unless, at some time in the future, investigators gained access to the inner workings of the mine and were able to gather enough evidence to prosecute.

There seems faint prospect of that. As this book was going to press, planning was underway for eventual re-entry into Pike's 2.3-kilometre stone tunnel, but only up to the point where it is known to be blocked by a massive roof collapse. First, the ventilation shaft must be filled with a mixture of grout and low-density concrete to block the flow of oxygen through the mine that could ignite another underground fire. Then the end of the tunnel must be sealed off by lowering a nozzle down a borehole and filling the cavity with expanding resin. Methane will then be drawn out of the tunnel via another borehole, with the aid of a vacuum pump.[26]

In early September 2013 the Solid Energy board and the government's High Hazards Unit – a beefed-up inspectorate established as a consequence of the Pike disaster – approved the plan, and the government allocated $7.2 million from the public purse to pay for it. After each of the complex tasks has been safely completed, the Mines Rescue men will enter and search the tunnel for the first time since November 19, 2010.

But there is, currently, no plan to go beyond the rockfall and into the places in the heart of the small mine where the men were last known to be working. The risks are thought to be too great and the costs unknown. Only if the mine is redeveloped at some time in the future will there be any effort to recover the bodies. And the mine will be redeveloped only if the price of coal rises enough to justify the investment that would be needed to turn Pike into a safe and viable commercial proposition.

Over the years that investors continued to fund Pike's grandiose predictions, the price of coal was high enough to send men day after day into a flawed and dangerous place. Now that they are dead, the price is not high enough to get them out.

On the West Coast, in Christchurch and Southland, in Australia, South Africa and the United Kingdom, families mourn. Half a world away in St Andrews, Scotland, Malcolm and Jane Campbell weep for their only son, Malcolm. He was 25 when he died in the Pike River mine. His voice was the last heard of the 29 men.

Malcolm had trained as a maintenance engineer, and at 21 had left the home he shared with his parents and younger sister Kerry to look for work and adventure in Australia. He'd driven trucks, and worked in wineries and in gold and uranium mines, often alongside his cousin and best friend, George Campbell. In 2008 he came to New Zealand, and ended up on the West Coast, where he was recruited by Pike. He loved the work, as he loved all of life. He was vibrant, funny and energetic, a young man who attracted new friends wherever he went. When Daniel Rockhouse got married, Malcolm was his best man.

'Everyone remembers him as being a nice laddie,' his father said. 'He was such a happy guy. He just wanted to learn and get better, and go up the ranks.' In time he hoped to return across the Tasman with his wife-to-be, Amanda Shields, and her young daughter, and work his way up the management hierarchy in a big Australian mine.

On December 18, 2010 – less than a month after the explosion that killed him – he would have married Amanda, the young West Coaster he had met and fallen in love with in Greymouth. He had been making preparations to mine near the property of one of his workmates for gold with which to make his and Amanda's wedding rings. His parents had their plane tickets to New Zealand booked.

Like the families of the other men, Malcolm, Jane and Kerry Campbell have struggled since November 19, 2010. Malcolm was unable for a long time to do his normal job as a quarry manager, because the sound of blasting would conjure images of his son meeting his death in a mine 18,000 kilometres away. Jane lost her son and both her parents within a period of 14 months.

'It's so very hard to think of our son still lying over there in New Zealand,' Malcolm said. 'I think, in the end, life will never be the same and we will never have closure until we have Malcolm home. But in reality I don't think we will ever get him home. We miss him dearly.'

They want to fight on, for accountability and for the recovery of the men's remains, but at the same time they know that their beloved boy would say to them, 'Mum and Dad, you must try to get on and live your life.'[27]

They take some comfort from the knowledge that their son and brother does not lie alone. Together, the Pike 29 remain entombed.

Timeline

1979
Otter Minerals Exploration, a subsidiary of Mineral Resources New Zealand, obtains two prospecting licences (later re-issued as one licence) to explore the Pike River coalfield.

1980-84
Terry Bates, exploration manager of Otter Minerals Exploration, carries out mapping and exploration of the Pike coalfield, assisted by Canterbury University geology students; six exploration boreholes are drilled.

1982
Pike River Coal Company Ltd is formed to hold the Pike prospecting rights.

1987
Paparoa National Park is formed, adjacent to and partly overlaying the Pike coalfield. It is accepted that development of the coalfield within the park boundaries can take place, provided there are significant economic benefits and adequate environmental safeguards.

1988
Pike River Coal Company passes into the hands of New Zealand Oil & Gas, with the Pike licence valued at $7.65 million. Over subsequent years NZOG unsuccessfully seeks a partner to help it develop the coalfield.

1997-98
New Zealand Oil & Gas and Pike River Coal Company move to Sydney; finance manager Gordon Ward assumes responsibility for pursuing resource consents and access agreements necessary to develop a mine at Pike. A pre-feasibility study for an underground mine at Pike is prepared by Minserv International.

2000
Australian consultancy Minarco prepares a final feasibility report confirming the viability of an underground mine at Pike and predicting a return on investment of 29 percent. In response, people associated with Minarco take a 25 percent shareholding in Pike River Coal Company and Minarco's Graeme Duncan joins the board.

2004
Minister of Conservation Chris Carter approves an access agreement allowing the mine to be developed. Pike River Coal Company and the Department of Conservation negotiate rules under which the mine can operate. A deal is signed on October 24. Later, many of the rules will be found restrictive and renegotiated. New Zealand Oil & Gas relocates back to Wellington, with Gordon Ward as general manager. The company decides to develop the mine itself under the aegis of Pike River Coal Company.

2005
Peter Whittall is recruited as mine manager, responsible for developing the mine. He and Gordon Ward establish the business case for the project: mine development will cost $124 million; the first coal will be produced by September 2006. Saurashtra Fuels invests $17 million in return for a shareholding of 10.6 percent, and Saurashtra representative Dipak Agarwalla joins the Pike River Coal Company board.

2006
Pike River Coal Company is renamed Pike River Coal Ltd. In April, retired investment banker Denis Wood becomes chair of the board, taking over from long-time incumbent Tony Radford, who remains a director. Gujarat NRE invests $20 million, and NZOG's ownership is diluted to 54 percent. Behre Dolbear Australia assures bankers the mine can achieve its forecast production tonnages. Peter Whittall tells a mining conference that Pike has 'medium to high' methane levels that will be 'difficult to control by ventilation means alone'. In September, construction begins on the 2.3-kilometre tunnel to the coal seam. In December Wood resigns, along with fellow directors Graeme Duncan and James Ogden.

2007
In January, Gordon Ward becomes chief executive. John Dow, a retired gold mining executive, becomes chair of the board in May. In June, there is an initial public offering (IPO), which raises $85 million in capital. According to the prospectus, the mine will be producing coal by March 2008, and by 2009 production will reach a million tonnes a year. The cost of development is put at $174 million and the coal seam is said to contain only 'low to moderate' levels of methane. In July, Arun Jagatramka joins the board, representing Gujarat NRE.

2008
Development costs are now put at $196 million. Construction of the access tunnel continues; it will ultimately take twice as long and cost twice as much as expected. The first coal exports are now forecast for the end of the year. The company raises a further $60 million from investors, and borrows $40 million from Liberty Harbor, a Goldman Sachs investment company. In October, Kobus Louw becomes statutory mine manager; he holds the position four months, resigns, and leaves. Peter Whittall announces the project has finally hit coal. In November, ten gas ignitions terrify workers. The mine is

officially opened with a ceremony on site. The Department of Conservation bestows an award on Pike River Coal for its efforts to marry modern mining with a high standard of environmental protection.

2009

In January, Minister of Conservation Tim Groser visits Pike and calls it a showcase of modern mining. In February, the lower section of the ventilation shaft collapses and the company goes back to investors for another $45 million. Nigel Slonker becomes statutory mine manager and resigns five months later. In-seam drilling shows that an unexpected band of rock (graben) more than 200 metres wide must be penetrated before the coal seam is reached and production can begin. In October, Gordon Ward tells an Australian Stock Exchange conference that Pike is a showcase project that will mine the world's lowest ash coking coal, and has the potential to mine deep into the Paparoa seams for a further eight million tonnes of hard coking coal. Also in October, the emergency exit up the ventilation shaft is trialled and found to be unusable.

2010

In February, Pike's first shipment of coal, 20,000 tonnes, leaves Lyttelton bound for India. In April, an explosion at the Upper Big Branch coal mine in West Virginia kills 29 coal miners out of 31 underground. Pike raises a further $50 million from shareholders, investors are told the mine will soon be producing a million tonnes of coal a year, and will generate $4 billion over its lifetime. The company strikes a deal for NZOG to take over the debt owed to Liberty Harbor. In July, Gordon Ward reveals that by the start of 2011 the company will be $5.8 million short of cash – even if it achieves its forecast production of 620,000 tonnes of coal. In September, Ward is dismissed and Peter Whittall is appointed chief executive. Pike's second shipment of coal, 22,000 tonnes, leaves Lyttelton for India. Extraction using the hydro monitor begins, but production levels are much lower than expected. In October, the main ventilation fan is installed underground. On November 19 an explosion occurs deep inside the mine. Two workers make it out but 29 others are trapped. Dangerous gas levels make it impossible to attempt a rescue. On November 24 a second explosion rocks the mine, followed by further explosions on November 26 and November 28. Prime Minister John Key announces a royal commission of inquiry into the tragedy. On December 2 a national memorial service for the 29 dead miners is held at Omoto Racecourse near Greymouth. On December 13, Pike River Coal Ltd is placed in receivership.

2011

On January 13, police announce they are abandoning attempts to recover the 29 bodies. The Royal Commission on the Pike River Coal Mine Tragedy begins public hearings in July. In November, the Department of Labour lays charges against Peter Whittall, Pike River Coal Ltd and in-seam drilling contractor Valley Longwall International Pty Ltd for breaches of the Health and Safety in Employment Act.

2012
On July 31, Valley Longwall pleads guilty to three charges; it is later fined $46,800. Pike River Coal says it will not fight the nine charges against it. Peter Whittall pleads not guilty to 12 charges under the Health and Safety in Employment Act. Also in July, the mine is sold by the receivers to the government-owned mining company Solid Energy for $7.5 million. In November, the Royal Commission's findings are released. Kate Wilkinson immediately resigns as minister of labour.

2013
In June, Judge Jane Farish rules that Peter Whittall's trial will be moved to Wellington. The trial will be held in 2014 and is expected to take several months. In July, Judge Farish fines Pike River Coal $760,000 and orders it to pay $3.4 million in reparations to the families of the Pike 29 and the two men who survived the explosion. The company's receivers say there is only enough money left to pay each family $5,000. Also in July, police announce that no one will be charged with manslaughter in connection with the deaths of the 29 men.

Glossary

Coal mining and coal geology terms, including terms specific to Pike River Mine

ABM20 Type of continuous mining machine that cuts, gathers and loads coal, with onboard bolters for installing roof and rib (side) support; ABM stands for Alpine Bolter Miner.

adit Entrance to an underground mine that is horizontal or almost horizontal.

Alimak raise Section of Pike River mine's ventilation shaft that was constructed to bypass a collapsed section of the main shaft.

auxiliary fan Moveable fan used to ventilate dead-end roadways underground.

brattice Fire-resistant, impervious cloth used to direct ventilation inside a mine and to construct temporary stoppings.

carbon monoxide (CO) Odourless, colourless gas; in coal mines it can be produced by spontaneous combustion of coal, explosions and the use of explosives.

coal measures Sedimentary strata containing coal seams.

coalification Geological process in which heat and pressure turn peat into coal.

continuous miner Machine for developing roadways in coal; capable of continuously loading the cut coal into the transport system – for example, flume, shuttle car or conveyor.

crib Lunch.

cross-cut Underground roadway to connect two main roadways.

deputy Trained and qualified mine official, usually a holder of a deputy statutory certificate of competency, who carries out safety inspections and makes daily reports required by law on gas levels and ventilation. Reports to an **underviewer**.

driftrunner Specialised mine vehicle used to transport miners in and out of a mine.

egress Exit from a mine.

flameproof Flameproof equipment is enclosed in special housing to ensure any ignition is safely contained.

Floxal Machine used to generate and pump nitrogen into a mine to make the atmosphere inert.

flume Open steel channel for transporting coal and water away from mining areas.

GAG (Górniczy Agregat Gaśniczy) Machine used to pump inert gases and water vapour into a mine to extinguish fire and stabilise the atmosphere after an explosion.

goaf Void created by coal extraction.

headings Roadways, generally driven parallel, to access different parts of the mine.

heating *See* **spontaneous combustion**

hydro mining Process of excavating coal with the use of high-pressure water and specialised equipment. Also referred to as hydraulic mining.

inbye Direction towards the coalface from any point of reference within a mine.

intake Underground roadway with uncontaminated air flowing through it.

methane (CH_4) Odourless, colourless gas formed as a by-product of coalification, and emitted when coal is mined. Methane is explosive within the range of five percent to 15 percent of air. Also referred to as coal seam gas.

New Zealand Mines Rescue Service Trained professionals and volunteers who respond to emergencies in mines and tunnels; administered by the Mines Rescue Trust and funded by industry levies.

outburst Sudden violent eruption of coal, gas and rock from a coalface.

outbye Direction away from coalface from any point of reference within a mine.

outcrop Rock and/or coal exposed on the surface.

panel Mining area consisting of access roads, development headings and/or extraction areas, and usually with a separate ventilation circuit.

pit bottom in coal At Pike River mine, area in the coal seam beyond the end of the main access tunnel; contained water storage, pumping systems, electrical infrastructure and main ventilation fan.

pit bottom in stone At Pike River mine, area partway up the main access tunnel; contained underground services for coal collection, crushing and transport, pumping systems and electrical infrastructure.

rank Coalification level achieved by coal; assessed using a variety of properties, including carbon content, energy and volatile matter.

return Underground roadways that have contaminated air moving through them towards the surface after the air has passed a mining area.

roadheader Purpose-built machine for driving roadways in stone or coal and capable of loading the cut material into a transport system – for example, flume, loader, conveyor.

self-rescuer Breathing device that provides 30 to 60 minutes of oxygen for a mine worker when air becomes unbreathable.

shotcrete Concrete sprayed at high velocity on to a surface.

shot-firing Use of explosives to dislodge coal from a coalface.

slimline shaft At Pike River mine, small shaft connecting underground workings to the surface. It was constructed to augment ventilation after the main ventilation shaft collapsed in early 2009.

spontaneous combustion (in mining) Process in which coal reacts with oxygen to create heat. If the heat that is liberated accumulates, the rate of the reaction increases and there is a further rise in temperature. If the process is not detected the coal will eventually start to burn, a process known as a heating.

statutory certificate of competence Certificate issued in accordance with the Health and Safety in Employment (Mining Administration) Regulations 1996 to a person who has met the specified training and experience requirements.

statutory mine manager Holder of a required certificate of competence, who carries legal responsibility for compliance with health and safety legislation, manages day-to-day activities, and carries out (or delegates to another person) statutory inspections, examinations and reporting.

stopping Permanent or temporary structure built across a roadway in a mine to direct airflow.

stub Small dead-end extension off a main roadway in a mine for storing equipment, or as a position from which to in-seam drill, or to enable two vehicles to pass each other.

tube bundle Gas sampling system consisting of a series of long tubes running to different parts of a mine and conveying samples of air to the surface for detailed analysis.

underviewer Supervisor responsible for planning and coordinating mine activities, ensuring safety systems are implemented and maintained, and carrying out inspections and examinations; holds an certificate of competence as a coal mine underviewer and reports to the statutory mine manager.

windblast High pressure displacement of air in a mine caused by sudden strata failure.

Endnotes

Prologue
1. Russell Smith, author interview.
2. Daniel Rockhouse, testimony to Royal Commission on the Pike River Coal Mine Tragedy; *Royal Commission on the Pike River Coal Mine Tragedy*, Volume 2, October 2012, p. 24.
3. Shane Bocock, author interview.
4. Dan Duggan, testimony to Royal Commission on the Pike River Coal Mine Tragedy.
5. Ibid.
6. Doug White, email to Garry McCure, November 19, 2010.
7. Mattheus Strydom, testimony to Royal Commission on the Pike River Coal Mine Tragedy.
8. Dan Duggan, author interview; transcript of Duggan's call to emergency services, November 19, 2010.
9. A memorial to all those known to have been killed in coal mining fatalities on the West Coast, unveiled in January 2013, lists 430 names.
10. Nan Dixon, author interview.
11. Anna Osborne, author interview.
12. *Royal Commission on the Pike River Coal Mine Tragedy*, Volume 2, October 2012, p. 228.

ONE Pike Dream
1. Dave Bennett, former exploration manager of New Zealand Oil & Gas Ltd, author interview.
2. The advantage of low ash coal is that it uses less energy and results in less waste in the process of manufacturing coke. Coke is an essential ingredient in the blast furnace steelmaking process and is formed when coal is heated at extremely high temperatures. Fluidity is a measure of how well the coal becomes fluid at high temperatures.
3. By the early 1990s the four key investment companies under Radford's control were United Resources Investment Holdings Ltd, Mineral Resources New Zealand Ltd (whose name was changed to Otter Gold Mines Ltd in 1995), New Zealand Oil & Gas Ltd, and Gold Resources Ltd. Radford chaired all four and they were tied together by interlocking shareholdings that made them resistant to takeovers.
4. Brian Gaynor, 'Investors cry for guidance', *The New Zealand Herald*, November 27, 1999.

5. Radford saw off repeated attempts by Sir Ron Brierley's Guinness Peat Group to get a seat on the board of United Resources Investment Holdings Ltd in the early 1990s. In 1999 the High Court ordered Radford removed as the chair of Otter Gold Mines Ltd after he had disallowed the votes of shareholder Guinness Peat Group.
6. Jane Newman, 'Paleoenvironments, coal properties and their interrelationship in Paparoa and selected Brunner coal measures on the West Coast of the South Island', unpublished PhD thesis, University of Canterbury, 1985.
7. Jane Newman, brief of evidence to Royal Commission on the Pike River Coal Mine Tragedy.
8. Dave Bennett, author interview.
9. Graham Mulligan, former general manager, commercial, of New Zealand Oil & Gas Ltd, and former director of New Zealand Oil & Gas Ltd and Pike River Coal Ltd, author interview.
10. This passage draws on the testimony of Masaoki Nishioka to the Royal Commission on the Pike River Coal Mine Tragedy and author interview with Roger O'Brien, former geologist at New Zealand Oil & Gas Ltd.
11. Masaoki Nishioka, email correspondence with author and testimony to Royal Commission on the Pike River Coal Mine Tragedy.
12. Masaoki Nishioka, testimony to Royal Commission on the Pike River Coal Mine Tragedy.
13. Dave Bennett, author interview.
14. Peter Liddle, former company secretary and director of New Zealand Oil & Gas Ltd and Pike River Coal Ltd, author interview.
15. Ibid.
16. Richard Inder, 'NZOG pages way to list Pike River', *National Business Review*, October 30, 1998.
17. 'Keen interest in Pike River coal', *The Press*, December 7, 1998.
18. Coal Marketing Services (Peter Gunn), *Pike River Coal Mine Prefeasibility Study*, prepared for New Zealand Oil & Gas Ltd, May 25, 1995.
19. Richard Inder, 'NZOG pages way to list Pike River', *National Business Review*, October 30, 1998.
20. Peter Gunn, email correspondence with author.
21. Minserv International Ltd, *Pre-Feasibility of the Pike River Coal Mining Project*, March 1998.
22. Minarco Pty Ltd was called AMC Resource Consultants Pty Ltd at the time the 2000 feasibility study was written; it changed its name to Minarco in 2001. In the interests of simplicity, it is referred to throughout the book as Minarco. Graeme Duncan resigned as managing director of Minarco in 2004 to focus on Pike River Coal Ltd. He continued to subcontract his services to Pike River Coal through his company Tasman Mining. David Meldrum, also a substantial shareholder in Pike, took over as managing director of Minarco.

23. AMC Resource Consultants Pty Ltd, *Final Feasibility Study for Pike River Mine*, Project Summary, 2000.
24. Graeme Duncan, director of Minarco Pty Ltd and former director of Pike River Coal Ltd, brief of evidence to Royal Commission on the Pike River Coal Mine Tragedy.
25. Ibid.
26. Graeme Duncan, author interview.
27. Peter Gunn, email correspondence with author.
28. Tony Frankham, former director of New Zealand Oil & Gas Ltd and Pike River Coal Ltd, former director of Otter Gold Mines Ltd and United Resources Investment Holdings Ltd, author interview.
29. Tony Kokshoorn, author interview.
30. Ibid.
31. Access Arrangement under Crown Minerals Act 1991, Mining Permit 41-453, Pike River Coal Ltd, between Chris Carter, Minister of Conservation, and Pike River Coal Company Ltd, October 24, 2004.
32. AMC Resource Consultants Pty Ltd, *Final Feasibility Study for Pike River Mine*, Project Summary, 2000.
33. Graeme Duncan, author interview.
34. Ibid.
35. Ibid.
36. Peter Whittall, testimony to Royal Commission on the Pike River Coal Mine Tragedy.
37. Ibid.

TWO Great Expectations

1. Peter Whittall and Gordon Ward, *Final Mine Plan and Financial Model*, report to Pike River Coal Ltd board of directors, July 13, 2005.
2. Mark Smith, Department of Conservation liaison officer, testimony to Royal Commission on the Pike River Coal Mine Tragedy.
3. Peter Whittall, 'Modern Mine Design', presentation to AusIMM conference, Auckland, New Zealand, November 2005.
4. Behre Dolbear Australia Pty Ltd *Pike River Coal Project – Greymouth, New Zealand: Pike River Production Comparisons*, report for Pike River Coal Ltd, signed by John McIntyre, June 30, 2006.
5. Denise Weir, author interview.
6. Pike River Coal Ltd, *New Zealand Prospectus*, May 22, 2007.
7. Gordon Ward, interview in *Sharechat*, November 30, 2004.
8. Pike River Coal Ltd, *New Zealand Prospectus*, May 22, 2007.
9. Ibid.
10. Behre Dolbear Australia Pty Ltd, *Pike River Coal Project – Greymouth, New Zealand: Independent Technical Review*, May 17, 2007.
11. Pike River Coal Ltd, *New Zealand Prospectus*, May 22, 2007.

12. Nick Churchouse, 'Pike River digs in after jump start', *The Dominion Post*, July 21, 2007.

THREE Early Warnings
1. Jane Newman, letter to Masaoki Nishioka and colleagues, November 25, 1992.
2. Jane Newman, letter to New Zealand Oil & Gas Ltd, November 21, 1991.
3. Peter Gunn, email to Jane Newman, August 2, 2001.
4. Jane Newman, email to Gordon Ward, October 3, 2001.
5. Jane Newman, email to Graeme Duncan, October 10, 2001.
6. Hugh Logan, author interview.
7. Department of Conservation, tier-two paper prepared for Royal Commission on the Pike River Coal Mine Tragedy, Appendix A.
8. Ibid.
9. Dr Murry Cave, technical evidence on behalf of Director-General of Conservation on resource consent applications by Pike River Coal Ltd to Grey District Council, West Coast Regional Council, Buller District Council, December 17, 2002.
10. Pike River Coal Ltd, *New Zealand Prospectus*, May 22, 2007.
11. Peter Whittall, 'Pike River Coal – Hydraulic Mine Design on New Zealand's West Coast', in Aziz, N (ed), *Coal 2006*, Coal Operators' Conference, University of Wollongong and Australasian Institute of Mining and Metallurgy, 2006.
12. Alan Wood, 'Geologist has advice for Pike River Coal investors', *The Press*, May 29, 2007.
13. Peter Whittall, letter to editor, *Greymouth Evening Star*, May 22, 2007.
14. Harry Bell, author interview.
15. Harry Bell, testimony to Royal Commission on the Pike River Coal Mine Tragedy.
16. Ibid.
17. Ibid.
18. Ibid.
19. Graeme Duncan, author interview.
20. Denis Wood, letter of resignation from Pike River Coal board of directors, December 7, 2006.
21. John Dow, testimony to Royal Commission on the Pike River Coal Mine Tragedy.
22. Ibid; Stuart Nattrass, author interview.

FOUR Trouble from the Start
1. Pike River Coal Ltd, *New Zealand Prospectus*, May 22, 2007.
2. Les Tredinnick, author interview.
3. Richard Cotton, author interview.
4. Les McCracken, email to Gordon Ward and Peter Whittall, October 22, 2008.
5. Les McCracken, author interview.

6. *Royal Commission on the Pike River Coal Mine Tragedy*, Volume 2, October 2012, p. 38.
7. Pike River Coal Ltd, press release, October 17, 2008.
8. Tony Kokshoorn, author interview.
9. 'Pike River Coal Mine West Coast NZ': http://www.youtube.com/watch?v=NT1-iaseARw
10. Tony Kokshoorn, author interview.
11. Russell Smith, author interview.
12. Gordon Ward, 'Pike River Coal – Making an Operating Coal Mine', paper for Australasian Institute of Mining and Metallurgy, August 2008.
13. Ibid.
14. Gordon Ward, address to Pike River Coal Ltd annual general meeting, November 28, 2008
15. Ibid.
16. Les Tredinnick, author interview.
17. Harry Bell, author interview; Harry Bell, testimony to Royal Commission on the Pike River Coal Mine Tragedy.
18. Kevin Poynter, email to Kobus Louw, November 14, 2008.
19. Harry Bell, testimony to Royal Commission on the Pike River Coal Mine Tragedy.
20. Kevin Poynter, email to Kobus Louw, November 19, 2008.
21. Kevin Poynter, email to Kobus Louw, December 24, 2008.
22. Kobus Louw, author interview.
23. Kobus Louw, email to Kevin Poynter, December 24, 2008.
24. Don McFarlane, brief of evidence to Royal Commission on the Pike River Coal Mine Tragedy; Evan Giles, author interview.
25. Joe Edwards, author interview.
26. Seth Tiddy, author interview.
27. Les Tredinnick, author interview.
28. Scott Campbell, author interview.
29. URS New Zealand, *Pike River Coal Mine: Ventilation Shaft Failure Model*, Final Report (Revised), prepared for McConnell Dowell, June 30, 2009.
30. Peter Whittall, testimony to Royal Commission on the Pike River Coal Mine Tragedy.
31. John Fisk, author interview.

FIVE Management Blues
1. Peter Whittall and Gordon Ward, *Final Mine Plan and Financial Model*, 2005; Pike River Coal Ltd, *New Zealand Prospectus*, May 22, 2007; Pike River Coal Ltd, *New Zealand Prospectus*, January 17, 2008; Pike River Coal Ltd, *Shortform Prospectus and Investment Statement*, March 16, 2009.
2. Pike River Coal Ltd, *New Zealand Prospectus*, January 17, 2008.
3. Pike River Coal Ltd, press release, February 19, 2009.

4. David Salisbury, author interview.
5. Russell Smith, author interview.
6. Kobus Louw, author interview.
7. Jonny McNee, author interview.
8. Jane Newman, author interview.
9. Jimmy Cory, email to Jane Newman, January 26, 2009.
10. Counsel for the families of men who died in Pike River Mine, final submissions to Royal Commission on the Pike River Coal Mine Tragedy.
11. Terry Moynihan, brief of evidence to Royal Commission on the Pike River Coal Mine Tragedy.
12. Catriona Bayliss, author interview.
13. Ibid.
14. Neville Rockhouse, testimony to Royal Commission on the Pike River Coal Mine Tragedy.
15. Denise Weir, author interview.
16. Les McCracken and Peter Whittall, email correspondence, April 3 and April 4, 2008.

SIX Many Whistles Blowing
1. Barry McIntosh, author interview.
2. Bernie Lambley, author interview.
3. Russell Smith, author interview.
4. Ibid.
5. Pike River Coal Ltd, *2010 Annual Review*.
6. Pike River Coal Ltd, *Quarterly Activities Report,* June 2009.
7. Barry McIntosh, author interview.
8. Marta Steeman, 'Pike River eyes shipping coal to shareholders', *The Press*, August 26, 2008.
9. Brent Mackinnon, author interview.
10. Nigel Slonker, author interview.
11. Les McCracken, author interview.
12. Ibid.
13. John Dow, author interview.
14. Dave Stewart, testimony to Royal Commission on the Pike River Coal Mine Tragedy.
15. John Dow, testimony to Royal Commission on the Pike River Coal Mine Tragedy.
16. Harry Bell, brief of evidence to Royal Commission on the Pike River Coal Mine Tragedy.
17. Simon Donaldson, Greg Fry, Peter Sattler, assessment papers, New Zealand Qualifications Authority Unit Standard 7142, September 2009.
18. Harry Bell, author interview.
19. Counsel for the families of men who died in Pike River Mine, final submissions to Royal Commission on the Pike River Coal Mine Tragedy.

20. *Royal Commission on the Pike River Coal Mine Tragedy*, Volume 2, October 2012, p. 199.
21. Access Arrangement under Crown Minerals Act 1991, Mining Permit 41-453, Pike River Coal Ltd, between Chris Carter, Minister of Conservation and Pike River Coal Company Ltd, October 24, 2004.
22. Terry Moynihan, brief of evidence to Royal Commission on the Pike River Coal Mine Tragedy.
23. New Zealand Mines Rescue Service, *Pike River Emergency Equipment and Self-escape Audit*, August 20, 2009.
24. Peter Whittall, testimony to Royal Commission on the Pike River Coal Mine Tragedy.
25. Adrian Couchman, testimony to Royal Commission on the Pike River Coal Mine Tragedy.
26. Gordon Ward, presentation to ASX Small to Mid Caps Conference, Hong Kong, October 29, 2009.
27. David Salisbury, brief of evidence to Royal Commission on the Pike River Coal Mine Tragedy.

SEVEN Too Big to Fail
1. David Salisbury, author interview.
2. David Salisbury, brief of evidence to Royal Commission on the Pike River Coal Mine Tragedy.
3. Pike River Coal Ltd, *2010 Annual Report*.
4. David Salisbury, brief of evidence to Royal Commission on the Pike River Coal Mine Tragedy.
5. Pike River Coal Ltd, *2010 Annual Report*.
6. Pike River Coal Ltd, presentation to Coaltrans Asia 2010, May 30–June 2, 2010.
7. McDouall Stuart Research, report to clients, May 5, 2010.
8. David Salisbury, brief of evidence to Royal Commission on the Pike River Coal Mine Tragedy.
9. New Zealand Oil & Gas Ltd, *2010 Annual Report*.
10. Behre Dolbear Australia Pty Ltd, *Pike River Coal Project – Greymouth, New Zealand: Independent Management Review*, May 19, 2010; Behre Dolbear Australia Pty Ltd, *Pike River Coal Project – Greymouth, New Zealand: Independent Technical Review*, May 20, 2010.
11. Dene Murphy, author interview.
12. Counsel for the families of men who died in Pike River Mine, final submissions to Royal Commission on the Pike River Coal Mine Tragedy.
13. Dene Murphy, author interview.
14. Ibid.
15. Ibid.
16. *Royal Commission on the Pike River Coal Mine Tragedy*, Volume 2, October 2012, p. 98.

17. Dene Murphy, author interview.
18. Alan Houlden, testimony to Royal Commission on the Pike River Coal Mine Tragedy.
19. John Dow, author interview.
20. Bernie Lambley, author interview.
21. Doug White, email to Dave Stewart, February 15, 2010.
22. Dave Stewart, *Pike River Compliance Audit*, Reports: February 11, 2010; March 3, 2010; March 10–11, 2010; March 18–19, 2010; March 31–April 1, 2010; April 8–9, 2010; April 15–16, 2010; April 23, 2010.
23. Dave Stewart, testimony to Royal Commission on the Pike River Coal Mine Tragedy.
24. Ibid.
25. Dene Murphy, author interview.
26. Ibid.
27. Alan Houlden, testimony to Royal Commission on the Pike River Coal Mine Tragedy.
28. Quintin Rawiri, author interview.
29. Bernie Lambley, author interview.
30. David Salisbury, brief of evidence to Royal Commission on the Pike River Coal Mine Tragedy.

EIGHT Marching to Calamity
1. Masaoki Nishioka, email correspondence with author.
2. Masaoki Nishioka, testimony to Royal Commission on the Pike River Coal Mine Tragedy.
3. Transcript, Royal Commission on the Pike River Coal Mine Tragedy, November 23, 2011; cross-examination of Masaoki Nishioka by John Haigh and Stacey Shortall.
4. 'Pike River Coal Bonuses', memo to staff, June 5, 2010.
5. Masaoki Nishioka, daily work record July–October 2010.
6. Masaoki Nishioka, testimony to Royal Commission on the Pike River Coal Mine Tragedy.
7. *Royal Commission on the Pike River Coal Mine Tragedy*, Volume 2, October 2012, p. 162–63.
8. Ibid, p. 163.
9. Masaoki Nishioka, daily work record July–October 2010.
10. Pike River Coal Ltd, press release, October 11, 2010.
11. *Royal Commission on the Pike River Coal Mine Tragedy*, Volume 2, October 2012, p. 84.
12. Masaoki Nishioka, daily work record July–October 2010.
13. *Royal Commission on the Pike River Coal Mine Tragedy*, Volume 2, October 2012, p. 93–94.
14. Masaoki Nishioka, email to Dave Perry, October 19, 2010.

15. Masaoki Nishioka, email to author, February 19, 2013.
16. George Mason, brief of evidence to Royal Commission on the Pike River Coal Mine Tragedy.
17. George Mason, testimony to Royal Commission on the Pike River Coal Mine Tragedy.
18. Stephen Wylie, brief of evidence to Royal Commission on the Pike River Coal Mine Tragedy.
19. Ibid.
20. Gordon Ward, police interview, September 29, 2011.
21. Peter Whittall, 'Modern Mine Design', presentation to AusIMM conference, Auckland, New Zealand, November 2005.
22. Brian Wishart, email to Jimmy Cory, April 11, 2010.
23. Drive Mining Pty Ltd, *Pike River Coal Ltd: Gas Drainage Assessment*, May 16, 2010.
24. Pieter Van Rooyen, testimony to Royal Commission on the Pike River Coal Mine Tragedy.
25. *Royal Commission on the Pike River Coal Mine Tragedy*, Volume 2, October 2012, p. 140.
26. Ibid.
27. *Royal Commission on the Pike River Coal Mine Tragedy*, Volume 2, October 2012, pp. 135–38.
28. 'Summary of the Reports of Certain Incidents and Accidents at the Pike River Coal Mine', November 2011, submitted in evidence to Royal Commission on the Pike River Coal Mine Tragedy.
29. *Royal Commission on the Pike River Coal Mine Tragedy*, Volume 2, October 2012, p. 143.
30. 'Summary of the Reports of Certain Incidents and Accidents at the Pike River Coal Mine', November 2011, submitted in evidence to Royal Commission on the Pike River Coal Mine Tragedy.
31. Gregory Borichevsky, brief of evidence to Royal Commission on the Pike River Coal Mine Tragedy.
32. Gregory Borichevsky, extract from police interview submitted in evidence to Royal Commission on the Pike River Coal Mine Tragedy.
33. *Royal Commission on the Pike River Coal Mine Tragedy*, Volume 2, October 2012, p. 200.
34. Adrian Couchman, testimony to Royal Commission on the Pike River Coal Mine Tragedy.
35. Ibid.
36. *Royal Commission on the Pike River Coal Mine Tragedy*, Volume 1, October 2012, p. 12.
37. Peter Whittall, address to Pike River Coal Ltd 2010 annual general meeting.
38. George Mason, email to Matt Coll, Terry Moynihan, Greg Borichevsky, Robb Ridl, Huw Parker, Steve Ellis, Doug White, October 31, 2010.

39. Doug White, email to Garry McCure, November 14, 2010.
40. Doug White, testimony to Royal Commission on the Pike River Coal Mine Tragedy.
41. Terry Moynihan, email to Doug White, November 10, 2010.
42. Andrew Harvey-Green, author interview.
43. Kim Joynson, brief of evidence to Royal Commission on the Pike River Coal Mine Tragedy.

NINE Who Will Say Stop?
1. Harry Bell, testimony to Royal Commission on the Pike River Coal Mine Tragedy; Harry Bell, author interview.
2. David Shanks and Jane Meares, *Pike River Tragedy: Report of the Independent Investigation to the Chief Executive of the Ministry of Business, Innovation and Employment*, March 2013.
3. Michael Firmin, testimony to Royal Commission on the Pike River Coal Mine Tragedy.
4. Ibid.
5. Ibid.
6. Kevin Poynter, testimony to Royal Commission on the Pike River Coal Mine Tragedy.
7. *Crush Incident – Pike River Coal*, Department of Labour investigation report, February 14, 2010.
8. Kevin Poynter, *Operational Review Process*, Monthly Report: June 2010, July 2010, September 2010, October 2010.
9. Mining Steering Group, Department of Labour, minutes of meeting, July 10, 2009.
10. David Shanks and Jane Meares, *Pike River Tragedy: Report of the Independent Investigation to the Chief Executive of the Ministry of Business, Innovation and Employment*, March 2013.
11. Ibid.
12. *Royal Commission on the Pike River Coal Mine Tragedy*, Volume 2, October 2012, pp. 199–200.
13. Kevin Poynter, testimony to Royal Commission on the Pike River Coal Mine Tragedy.
14. John Dow, email to author, July 8, 2013.
15. Pike River Coal Ltd, minutes of board meeting, November 15, 2010.
16. Doug White, email to Allen McFadzen, November 16, 2010.
17. John Dow, testimony to Royal Commission on the Pike River Coal Mine Tragedy.
18. Pike River Coal Ltd, *Annual Review*, 2010.
19. Michelle Gillman, brief of evidence to Royal Commission on the Pike River Coal Mine Tragedy.
20. Neville Rockhouse, testimony to Royal Commission on the Pike River Coal Mine Tragedy.

21. Ibid.
22. Ibid.
23. John Dow, testimony to Royal Commission on the Pike River Coal Mine Tragedy.
24. Hawcroft Consulting International, *Pike River Mine: Risk Survey, Underground, CPP and Surface Operations*, commissioned by Colin Whyte, associate director, Marsh Ltd New Zealand, Final Report, July 2010.
25. John Dow, testimony to Royal Commission on the Pike River Coal Mine Tragedy.
26. Doug White, testimony to Royal Commission on the Pike River Coal Mine Tragedy.
27. Ibid.
28. New Zealand Amalgamated Engineering, Printing and Manufacturing Union, final submissions to Royal Commission on the Pike River Coal Mine Tragedy.
29. Matt Winter, brief of evidence to Royal Commission on the Pike River Coal Mine Tragedy.
30. Ibid.
31. Neville Rockhouse, testimony to Royal Commission on the Pike River Coal Mine Tragedy.

TEN November 19
1. Pike River Coal Ltd, *Deputy Statutory Report*, November 8, 2010.
2. Anna Osborne, author interview.
3. Dene Murphy, author interview.
4. Scott Campbell, author interview.
5. *Royal Commission on the Pike River Coal Mine Tragedy*, Volume 2, October 2012, p. 21.
6. Ibid.
7. 'Summary of the Reports of Certain Incidents and Accidents at the Pike River Coal Mine', November 2011, submitted in evidence to Royal Commission on the Pike River Coal Mine Tragedy.
8. *Royal Commission on the Pike River Coal Mine Tragedy*, Volume 2, October 2012, p. 21.
9. Ibid.
10. Ibid.
11. Kevan Curtis, author interview.
12. Steve Ellis, testimony to Royal Commission on the Pike River Coal Mine Tragedy.
13. Ibid.
14. Ibid.
15. Steve Ellis, testimony to Royal Commission on the Pike River Coal Mine Tragedy.
16. Tony Kokshoorn, author interview.

17. Anna Osborne, author interview.
18. Dave Cross, author interview.
19. Trevor Watts, author interview.
20. Rob Smith, author interview.
21. Tony Kokshoorn, author interview.
22. Troy Stewart, author interview.
23. New Zealand Fire Service, Operation Pike incident log, November 19–25, 2010.
24. Dave Cross, author interview.
25. John Key, author interview.
26. Peter Whittall, brief of evidence to Royal Commission on the Pike River Coal Mine Tragedy.
27. Brian Small, Busby Ramshaw Grice, email to author, July 18, 2013.

ELEVEN Five Days
1. Lauryn Marden, witness statement to Royal Commission on the Pike River Coal Mine Tragedy; Lauryn Marden, author interview.
2. Tara Kennedy, brief of evidence to Royal Commission on the Pike River Coal Mine Tragedy.
3. Lauryn Marden, brief of evidence to Royal Commission on the Pike River Coal Mine Tragedy.
4. Laura Mills, 'Mine nightmare', *Greymouth Star*, November 20, 2010.
5. Laura Mills, 'Everything humanly possible being done', *Greymouth Star*, November 20, 2010.
6. Robin Hughes, author interview.
7. Trevor Watts, author interview.
8. Robin Hughes, testimony to Royal Commission on the Pike River Coal Mine Tragedy.
9. Robin Hughes, author interview.
10. New Zealand Fire Service, Operation Pike incident log, November 19–25, 2010.
11. John Key, author interview.
12. Ibid.
13. Darren Brady, author interview.
14. Craig Smith, testimony to Royal Commission on the Pike River Coal Mine Tragedy.
15. Steve Ellis, testimony to Royal Commission on the Pike River Coal Mine Tragedy.
16. Ibid.
17. Tara Kennedy, brief of evidence to Royal Commission on the Pike River Coal Mine Tragedy.
18. Bernie Monk, brief of evidence to Royal Commission on the Pike River Coal Mine Tragedy.
19. Ibid.
20. Carol Rose, brief of evidence to Royal Commission on the Pike River Coal Mine Tragedy.

21. Gerry Morris and Tony Kokshoorn, author interview.
22. New Zealand Fire Service, Operation Pike incident log, November 19–25, 2010.
23. Police National Headquarters, situation report, November 21, 2010; extract submitted in evidence to Royal Commission on the Pike River Coal Mine Tragedy.
24. Peter Whittall, testimony to Royal Commission on the Pike River Coal Mine Tragedy.
25. Gary Knowles, brief of evidence to Royal Commission on the Pike River Coal Mine Tragedy.
26. Grant Nicholls, testimony to Royal Commission on the Pike River Coal Mine Tragedy.
27. Catriona Bayliss, author interview.
28. John Key, author interview.
29. Laura Mills, 'Prayer and Despair', *Greymouth Star*, November 22, 2010.
30. Anna Leask, 'Hoping for best while preparing for worst', *The New Zealand Herald*, November 23, 2010.
31. Tui Bromley, 'Gloom descends', *Greymouth Star*, November 24, 2010.
32. Catherine Masters, 'Bleaker by the day', *The New Zealand Herald*, November 24, 2010.
33. Ibid.
34. Ibid.
35. Peter Whittall, brief of evidence to Royal Commission on the Pike River Coal Mine Tragedy.
36. Gary Knowles, author interview.
37. John Key, author interview.
38. Troy Stewart, author interview.
39. Peter Whittall, author interview.
40. Gary Knowles, brief of evidence to Royal Commission on the Pike River Coal Mine Tragedy.
41. Trevor Watts, author interview; Trevor Watts, brief of evidence to Royal Commission on the Pike River Coal Mine Tragedy.
42. Dave Cliff, author interview.
43. John Taylor, brief of evidence to Royal Commission on the Pike River Coal Mine Tragedy.
44. There is disagreement as to whether Knowles or Whittall stated there had been a second explosion. The portrayal of events in this paragraph draws on the recollections of Gary Knowles and Barbara Dunn.

TWELVE Entombed

1. Andrea Vance, 'Don't seal the mine, beg relatives', *The Dominion Post*, November 26, 2010.
2. Stuart Nattrass, author interview.
3. Doug White, testimony to Royal Commission on the Pike River Coal Mine Tragedy.

4. Laura Mills, 'Mine inferno', *Greymouth Star*, November 30, 2010.
5. Viv Logie, 'Decision on mine's future close – Pike River chairman', *Greymouth Star*, December 3, 2010.
6. Trevor Watts, brief of evidence to Royal Commission on the Pike River Coal Mine Tragedy.
7. Steve Ellis, testimony to Royal Commission on the Pike River Coal Mine Tragedy.
8. Wayne Hartley, general manager of Queensland Mines Rescue Service, letter to Howard Broad, New Zealand police commissioner, January 12, 2011; Gary Knowles, author interview.
9. Commissioner Howard Broad, file note, January 17, 2011.
10. Safety managers, responses to Commissioner Howard Broad in 'New Zealand Police – Operation Pike', report by David Reece, January 7, 2011.
11. Commissioner Howard Broad, file note, January 17, 2011.
12. Jim Stuart-Black, email to Catherine Petrey, Geraint Emrys, Paula Beever, Dave Cliff, David Bell, Wayne Hartley, Mike Munnelly, January 10, 2011.
13. Commissioner Howard Broad, file note, January 17, 2011.
14. Gary Knowles, letter to Commissioner Howard Broad, January 18, 2011.
15. Steve Ellis, testimony to Royal Commission on the Pike River Coal Mine Tragedy.
16. Kurt Bayer and Hayden Donnell, 'Pike River mine sold to Solid Energy', *The New Zealand Herald,* March 9, 2012.
17. Rebecca Macfie, 'The Pike River families have their hopes dashed again', *listener.co.nz*, May 11, 2012.
18. *Royal Commission on the Pike River Coal Mine Tragedy*, Volume 2, October 2012, p. 177.
19. *Royal Commission on the Pike River Coal Mine Tragedy*, Volume 1, October 2012, p. 12.
20. Ibid.
21. *Royal Commission on the Pike River Coal Mine Tragedy*, Volume 2, October 2012, p. 185–90.
22. Judge J. A. Farish, *Department of Labour v Pike River Coal Ltd*, judgement, May 3, 2013.
23. New Zealand Oil & Gas Ltd, statement to author.
24. Peter Read, author interview.
25. Ibid.
26. Trevor Watts, author interview; Solid Energy Ltd, statement to author.
27. Malcolm Campbell, author interview.

Index

ABM20 continuous miner 15, 133–34, 148, 179, 185, 186, 237
access tunnel 11–14, 16–17, 40, 43, 65–68, 105, 190, 191, 197, 220, 232–33, 242, 246
acid mine drainage (AMD) 50–52, 53, 56, 76
Adams, Conrad 11, 12, 21, 188
Agarwalla, Dipak 6, 129
Air New Zealand 200, 227
Alimak raise 76, 81, 99, 105, 112–13
atmospheric sampling (mine) 193, 195, 196, 198, 203–5, 207, 218–19

Bates, Terry 6, 24–25, 28, 47, 48, 53, 245
Bayliss, Catriona 93, 211, 213
B crew 186–87
Behre Dolbear Australia Pty Ltd 6, 42, 45, 121–22, 133, 152, 246
Bell, Harry 58–60, 72–73, 110, 111, 167–68, 169, 170, 179
Bell, Stewart 237
Bisphan, Craig 187
Bocock, Shane 13
bodies, recovery of 223, 224–25, 226, 230–31, 232, 233, 247
Borichevsky, Greg 157, 158, 172, 179
Brady, Darren 208, 211, 218, 241
breathing apparatus 12, 13, 17, 113, 130, 158, 208, 230, 232
Broad, Howard 230–31
Brown, Miles 152–53
Brownlee, Gerry 70, 207, 221, 222, 231

Busby Ramshaw Grice 200, 210–11
Butler, David 228

Campbell, Glen 220–21
Campbell, Jane 243, 244
Campbell, Kerry 243, 244
Campbell, Malcolm 15, 16, 21, 187, 190, 243, 244
Campbell, Malcolm (father of miner Malcolm Campbell) 243, 244
Campbell, Scott 81, 186, 187
Canterbury Coal Research Group 48, 49
capital raising *see under* Pike River Coal Ltd
carbon monoxide 12, 14, 127, 184, 192, 195, 198, 204–5, 208
Carter, Chris 33, 246
cave-ins 142, 158–59, 178, 207, 226, 232, 237, 241, 242
Cave, Murry 52–53, 55–56, 57–58
C crew 12, 15, 187–89
  see also individual names
Clayton, Miranda 200
Cliff, Dave 218, 219
coal seams, Pike River 23, 25
  Brunner 48, 71, 90
  exploration and mapping 25, 26, 27, 40, 47–50, 50–52, 53, 55, 56, 67, 88–89, 90, 101, 245; *see also* in-seam drilling
  Paparoa 48, 70–71
  see also geology
Coll, Matt 112–13

267

coking coal 23–24, 27, 29, 30, 39, 41, 44, 59–60, 69, 90–91, 101–2, 114–15, 120, 164, 224, 233, 247
continuous mining 40
　ABM20 continuous miner 15, 133–34, 148, 179, 185, 186, 237
　Waratah continuous miner 86, 92, 99–100, 103–4, 105–6, 121, 128, 132–33, 134, 160, 187
contractors 34–35, 43–44, 141, 165, 188, 224, 228–29
　CYB Construction 188, 201
　Ferguson Brothers 43–44
　Graeme Pizzato Contracting 185, 188
　McConnell Dowell 8, 9, 59, 65, 67, 68, 72, 75, 76, 78, 79, 81–82, 151, 169, 186, 189
　Skevington Contracting 189, 190
　SubTech 20, 185, 202, 224
　Valley Longwall International 9, 101, 127, 156, 160, 187, 238, 240, 247, 248
control room 13, 14, 15, 17, 102, 130, 153, 154–55, 179, 190–91, 192, 197, 214, 220, 235
Cory, Jimmy 92, 151
Cotton, Richard 47–48, 67
Couchman, Adrian 113–14, 157–58, 159
criminal charges 238–39, 242, 247–48
Cross, Dave 192, 194, 198
Crown Minerals 168
Cruse, Glenn 21, 188
Curtis, Kevan 188–89, 190

Daly, Tom 193
Department of Conservation (DOC) 25, 29, 31–32, 33, 40, 50, 52, 55, 56, 76, 89, 98, 99, 112, 139, 246, 247

Department of Labour 169, 170–71, 208, 213, 225, 226, 236, 237, 238, 240, 247
　Health and Safety in Employment Act 73, 169, 196, 231, 238, 242, 247, 248
　inspectors/inspection 7, 8, 73, 158, 169–71, 170–73, 205, 213, 236
Dixon, Allan 15, 17, 18–19, 21, 81, 112, 183, 184, 186
Dixon, Bob 142
Dixon, Jack 18
Dixon, Nan 17–19, 234–35
Donaldson, Simon 110, 184
Dow, John 6, 58, 62, 63, 107–8, 129, 134–35, 140, 149, 162, 173–74, 175, 176, 177, 226, 234, 246
Drew (Verhoeven), Zen 21, 188
drift *see* access tunnel
drilling *see* coal seams, Pike River, exploration; *and* in-seam drilling
Duggan, Chris 21, 187
Duggan, Dan 13, 14–16, 190–91, 191–92
Dunbar, Joseph 21, 186–87
Duncan, Graeme 7, 30–31, 33, 35, 51–52, 56, 61–62, 245, 246
Dunn, Barbara 202–3, 220, 221
Durbridge, Rick (Rowdy) 188

Edwards, Joe 76, 79
egress, secondary *see* emergency exits
Elder, Don 233, 234
Ellis, Steve 7, 157, 158, 159, 179, 184, 190–91, 192, 205, 206, 208–9, 219, 220, 228, 232–33, 234
emergency exits 41, 55, 75, 86, 105, 110, 112, 113–14, 138, 140, 141, 157–58, 166, 172–73, 179–80, 203, 235

Engineering Printing and
Manufacturing Union (EPMU)
180–81
environmental management 25, 40, 46,
50, 56, 70, 98–99, 245, 247
escape (Rockhouse/Smith) 12–14, 17,
192, 195, 203, 208
explosions
  atmospheric sampling 193, 195, 196,
  198, 203–5, 207, 218–19
  cause 237–38, 240–42
  fire (post-explosions) 203–5, 206,
  210, 211, 219, 224–26
  first explosion 11–17, 190–219, 247
  fourth explosion 21, 225, 226, 247
  mine re-entry plans 212–13, 218–19,
  230–31, 232–33, 234, 242–43
  mine stabilisation (post-explosions)
  224–25, 226, 229–31, 232–33
  rescue operation 194–98, 203,
  207–8, 211, 212–13, 214–17,
  218–19, 220
  sealing the mine 206, 207–8, 211,
  217, 231, 232
  second explosion 21, 219–24, 247
  third explosion 21, 225, 247
  video footage 197, 203, 205, 208–9,
  213–14, 214–15, 216, 217, 220, 232

Farish, Jane (Judge) 238–39, 248
families (of Pike 29) 198, 199, 200,
201–3, 206, 207, 209–10, 211, 212,
214–15, 221–22, 223, 229, 231–32,
233, 234, 238, 243–44, 248
see also individual names
Firmin, Michael 7, 169–70, 173
First New Zealand Capital 60, 61, 62
Fisk, John 227, 230
Fry, Greg 110–11

GAG (Górniczy Agregat Gaśniczy)
224–25, 229–30
gas drainage *see* methane, drainage
gas management *see* methane,
management; *and* ventilation
geology 23, 26, 47–48, 50–52, 53, 54,
57, 66–67, 88–89, 90–91, 101–2
  see also coal seams, Pike River;
  Hawera Fault; *and* methane
Giles, Evan 78, 87
glossary of terms 249–51
Goodwin, Tony 7, 39, 86, 92–93, 99,
144
graben 101–2, 103, 127, 139
Grey District Council 33, 56, 228
Greymouth 37–38, 43, 209, 212, 226
Greymouth Star 203, 215, 226
Gribble, Nick 7, 104, 113–14, 157–58
Groser, Tim 99, 247
Gujarat NRE 7, 41, 61, 149, 175,
227–28, 246
Gunn, Peter 7, 29, 31, 50, 51, 52, 56,
107

Haddock, Peter 20, 193, 228
Hale, John 21, 188, 189
Harvey-Green, Andrew 162–63
Hawcroft Consulting 177–78
Hāwera Fault 23, 55, 57, 68, 139
health and safety 71, 93, 109–14,
122–27, 130, 131–32, 158, 162, 163,
169–73, 174–81, 186, 196, 235, 236,
238–39
Health and Safety in Employment Act
73, 169, 196, 231, 238, 247, 248
  charges under 238–39, 242
Henry, David 237
Herk, Daniel (Dan) 12, 21, 180, 188
Hoggart, David (Dave) 15, 21, 187

Holling, Richard (Rolls) 21, 187–88
Horne, Angela 134, 199
Houlden, Alan 122, 125–27, 131–32
Hughes, Robin 203–6, 207, 211, 226
Huntly mine 59
Hurren, Andrew (Huck) 21, 188
hydro mining 15, 26, 40, 91, 99, 106, 135, 137–43, 146–48, 150, 155, 157, 158–59, 160–61, 162, 172, 178, 183–84, 190, 235, 247
    bonus 141–42, 157

in-seam drilling 9, 89–90, 101, 127, 150–51, 152–53, 156, 160, 185, 186, 235, 237, 238, 240, 241, 247, 247
inspection (mine) *see* mining inspectorate
international standards 54–55, 124, 128, 144, 162, 174, 179–80
    see also regulations

Jagatramka, Arun 7, 8, 175, 246
Jamieson, Dean 153–54
Jonker, Koos 21, 187
Joynson, Kim 164, 165–66, 239
Joynson, Willie 21, 164–66, 181, 187, 239
Juganaut loaders 101, 121

Keane, Riki (Rik) 12, 21, 185, 189
Kennedy, Dave 24, 28
Kennedy, Tara 202, 209
Key, John 207, 214–15, 216, 217, 247
Kidd, John 120
King, Greg 228
Kitchin, Terry 22, 185, 202
Knapp, Dick 7, 123, 180, 199
Knowles, Gary 202–3, 206, 210, 212, 213, 215, 216, 218–19, 220, 221–22, 231

Kokshoorn, Tony 32–33, 69, 192–93, 198, 199, 200, 202, 210–11, 215–16, 227

Lambley, Bernie 7, 131, 133
legislation *see* international standards; *and* regulations
Lerch, Mick 7, 129–30, 144
Liberty Harbor 83–84, 103, 119, 120, 246, 247
licences
    exploration 24–25, 25–26, 29, 48, 49, 245
    mining 27, 32, 34, 90, 117, 168
Liddell, Ivan 7, 38, 89, 98
Liddle, Peter 28
Logan, Hugh 52
Louw, Kobus 8, 73–74, 76–77, 80, 85–87, 92, 105, 124, 177, 246
Loyalka, Sanjay 8

Mackie, Beth 239
Mackie, Samuel (Sam) 22, 185, 239
Mackinnon, Brent 103–5
MacLean, Neil (Judge) 232
Marden, Francis 22, 188, 201
Marden, Lauryn 201–2
Mason, George 147, 148, 160, 190
McBride, Paul 226
McConnell Dowell 8, 9, 59, 65, 67, 68, 72, 75, 76, 78, 79, 81–82, 151, 169, 186, 189
McCracken, Les 8, 68, 76, 94–95, 106–8, 128–29, 177
McCure, Garry 16, 161, 162
McIntosh, Barry 97–98, 102, 155, 220
McIntyre, John 42, 121
McKenzie, Lance 185
McLean, Bruce 234
McMorran, Tim 67

McNaughton, Callum 188–89, 190
McNee, Jonny 59, 87–92
media coverage 193, 199, 203, 210–11, 215, 216, 226
memorial service 226–27, 247
methane 27, 57, 184
   detection 98, 127, 130, 153, 154, 155–56, 179, 185, 235, 237, 240
   drainage 54–55, 57, 127, 150–53, 203
   emissions 16, 21, 45, 53–54, 86, 91, 139, 141–42, 143, 146, 150–51, 155, 157, 159–60, 172, 174–75, 185, 187, 197, 236, 237
   gassy mines 27, 54–55, 122–23, 167, 196
   ignitions 54, 71–74, 109, 111, 123, 170, 180, 197, 237, 246
   management 27, 45, 110, 124–25, 130, 178, 235–36
   outbursts 54, 55, 122–23, 152, 153
   see also explosions; in-seam drilling; *and* ventilation
Meyer, Ray 8, 61–62, 63, 118
Minarco 7, 30, 34, 50, 51, 53, 55, 245
mine layout *see illustration section*
   access tunnel 11–14, 16–17, 40, 43, 65–68, 105, 190, 191, 197, 220, 232–33, 242, 246
   Alimak raise 76, 81, 99, 105, 112–13
   pit bottom 81
   pit bottom in coal 57, 101, 139
   pit bottom in stone 12, 68, 190, 192
   slimline shaft 105, 114, 196, 213, 218, 219, 232, 241
   Spaghetti Junction 12, 125, 145, 151, 189
   Thunder Dome 125
   ventilation shaft (main) 75–82, 112, 155, 158, 203–4, 225, 242, 247

Mineral Resources New Zealand Ltd 6, 24–25, 245
mine re-entry plans 212–13, 218–19, 230–31, 232–33, 234, 242–43
mine stabilsation (post-explosions) 224–25, 226, 229–31, 232–33
mining disasters 18–19, 37, 147, 151, 167, 171, 194, 197, 199, 247
   Brunner Mine 18, 22
   Mt Davy mine 54, 122
   Strongman Mine 18, 19, 58, 72
mining inspectorate 58, 59, 71, 72, 73, 74–75, 110–11, 112, 124, 158, 163, 167–73, 242
Mitsui Mining 26, 29, 49
Monk, Bernie 209, 229, 231–32, 233, 234
Monk, Kath 209–10, 239
Monk, Michael 22, 188, 209
Morris, Gerry 210, 228
Moynihan, Terry 112, 162
Mt Davy mine 54, 122
Mudge, Stuart (Stu) 22, 188, 210
Mulligan, Graham 26, 28
Murphy, Dene (Mad Dog) 122–25, 131, 145, 185

Nattrass, Stuart 8, 62–63, 224
Newman, Jane 25, 47–50, 50–52, 88, 90, 92
Newman, Nigel 25, 49
New Zealand Fire Service 198, 206, 211, 231
New Zealand Mines Rescue Service 15–16, 17, 58, 113, 191–92, 194–97, 198, 203–5, 207–8, 213, 214, 217–18, 219, 220–21, 225, 226, 232, 242
New Zealand Oil & Gas 6, 8, 9, 26, 27–31, 34–35, 39, 41, 48–50, 61–62, 84, 90, 115, 117–19, 120–21, 122,

New Zealand Oil & Gas (*continued*)
134–35, 149, 154, 163–64, 168,
223–24, 227–29, 239–40, 245, 246,
247
New Zealand Police 195, 211, 214, 216,
217, 225, 226, 229–30, 231–32,
240–42, 247, 248
  Cross, Dave 192, 194, 198, 212–13
  Dunn, Barbara 202–3, 220, 221
  inquiry 240–42
  investigation 216
  Knowles, Gary 202–3, 206, 210,
212, 213, 215, 216, 218–19, 220,
221–22, 231
  Nicholls, Grant 213, 218
  Read, Peter 240–41, 242
Nieper, Kane 22, 188
Nishioka, Masaoki (Oki) 26–27, 49,
135, 137–43, 144, 145–146, 147, 150,
155, 162
Nolan, Mike 196

O'Brien, Roger 26, 27–28, 29
O'Neill, Peter 15, 19, 22, 183, 184, 186
OceanaGold 38, 52
Ogden, James 60, 61, 246
Osborne, Anna 20–21, 184–85, 193,
239
Osborne, Milton (Milt) 20, 21, 22,
184–85, 189, 193, 224
Otter Minerals Exploration 6, 24–25,
245
outbursts *see under* methane

Palmer, Brendon 22, 188
Panckhurst, Graham (Justice) 237
Paparoa National Park 25, 37, 245
Pike River Coal Ltd
  annual general meetings 70–71,
160–61

audits 128–31, 175
board of directors 6, 7, 8, 9, 39, 41,
60–63, 87, 121, 127–28, 173–76,
224, 236, 239, 240, 246
capital raising 23–24, 26, 28–32,
39, 41–42, 44–46, 60, 61, 83–84,
94–95, 119, 120–121, 122, 134,
149–50, 154, 163–64, 174, 199,
223–24, 246, 247
charges against 238–39, 247
coal processing plant 91, 106, 121,
146
environmental management 40, 46,
50, 56, 70, 98–99, 245, 247
estimates (production) 27, 29–30,
34, 39, 42, 44, 57, 70, 90, 94–95,
102, 127, 134, 148–49, 150, 160,
162, 236
feasibility studies 29–30, 31, 50–52,
53, 55, 56
health and safety 71, 93, 109–14,
122–27, 130, 131–32, 158, 162,
163, 169–73, 174–81, 186, 196,
235, 236, 238–39
licences (exploration/mining)
24–25, 25–26, 27, 29, 32, 34, 48,
49, 90, 117, 168, 245
management 41, 87, 93, 98, 106–7,
120–21, 122, 128–29, 135, 161,
177, 178, 181; *see also* statutory
mine manager
opening ceremony 70, 247
production levels 70, 82, 115, 127,
150, 160, 163, 247
receivership 227–29, 231, 232, 233,
247
reparation 239–40, 248
resource consents 32, 33, 34, 56, 245
risk assessment and management
45, 50–51, 53, 54–55, 56–57,
78–79, 97–98, 113, 114, 141–42,

144–45, 152, 174, 175, 176,
  177–78, 230, 236
Royal Commission 235–36, 238–40
sale to Solid Energy 233, 248
sharemarket listing 29, 42–43,
  44–46, 56–57, 60, 61, 62, 115,
  120–21, 160, 161–62, 246
police *see* New Zealand Police
Poynter, Kevin 8, 73–74, 74–75, 158,
  170–73
prospecting licence *see* licences,
  exploration

Queensland's Safety in Mines Testing
  and Research Station (SIMTARS)
  207, 208, 211

Radford, Tony 8, 23, 24–25, 27–28, 31,
  33, 39, 48, 60, 63, 117, 118, 148, 149,
  173, 234, 246
Rawiri, Quintin 132–34
Rawson, Stephen 8
Read, Peter 240–41, 242
Reece, David 230, 231, 237, 240
re-entry *see* mine re-entry plans
regulations 72–73, 109–10, 111, 112,
  124, 153, 156, 158, 167–69, 170,
  172, 173, 181, 208, 236, 242, 251
  see also international standards
Renk, Udo 8, 92
rescue operation 194–98, 203, 207–8,
  211, 212–13, 214–17, 218–19, 220
Ridl, Robb 8, 191, 199
risk assessment and management 45,
  50–51, 53, 54–55, 56–57, 78–79,
  97–98, 113, 114, 141–42, 144–45,
  152, 174, 175, 176, 177–78, 213,
  218–19, 230, 236
  see also health and safety
roadheader 15, 43, 72, 74, 81, 86, 92,
  101, 109, 134, 141, 187, 250

Rockhouse, Benjamin (Ben) 22, 187
Rockhouse, Daniel 12–14, 16, 17, 21,
  188, 190, 192, 195, 208, 209, 239
Rockhouse, Neville 8, 14, 93, 113, 114,
  159, 175–77, 181
Rodger, Peter (Pete) 22, 188
roof collapse *see* cave-ins
Rose, Carol 210, 229, 234
Rose, Stephen 210
Royal Commission on the Pike River
  Mine Tragedy 234–38, 247
Rūnanga 18

safety *see* health and safety;
  regulations; *and* risk assessment and
  management
Salisbury, David 9, 119, 120–21, 135,
  149
Sattler, Peter 108–9, 110
Saurashtra Fuels 6, 41, 61, 227–28, 246
sealing the mine 206, 207–8, 211, 217,
  231, 232
Seiko Mining 137, 138
self-rescuers *see* breathing apparatus
Shields, Amanda 243
Sims, Blair 15, 21, 22, 187, 189, 227
Sims, Lynne 21
Sims, Trevor 21
slimline shaft 105, 114, 196, 213, 218,
  219, 232, 241
Slonker, Nigel 9, 102–3, 105–6, 247
Smith, Colin 229
Smith, Craig 205–6
Smith, Denis 188
Smith, Rob 194–97
Smith, Russell (surviving miner)
  11–12, 13–14, 16, 69, 84–85, 101,
  189, 191, 192, 195, 208, 213, 216, 239
Smith, Russell (Mines Rescue
  brigadesman) 196–97

Solid Energy 38, 42, 50, 56, 137, 233–34, 242, 248
spontaneous combustion 130, 184, 241
Spring Creek mine 38, 42, 56, 137, 233
statutory mine manager 7, 8, 9, 73, 85, 102–3, 111, 124, 144, 157, 177, 178, 196, 205, 228, 232–33, 235, 246, 247,
Stewart, Dave 29–30, 39, 107, 108, 128–31, 154, 177, 197, 225
Stewart, Jolene 217, 218, 220
Stewart, Troy 194–95, 197, 217–18, 220–21
stone tunnel *see* access tunnel
Strongman Mine 26, 88, 135, 137
   disaster 18, 19, 58, 72
Strydom, Mattheus 16–17, 191
Stuart-Black, Jim 231
SubTech 20, 185, 202, 224
survivors *see* Rockhouse, Daniel; *and* Smith, Russell

Taylor, John 219
Thompson, Nick 211, 218, 220, 221
Tiddy, Seth 80
Titan Mining 132–34
training 92, 125–26, 129, 131, 132, 138, 147, 148, 153–54, 161, 165, 179, 236
   see also Rockhouse, Neville
Tredinnick, Les 65–66, 68, 69, 71–72, 73, 80, 151, 189

Ufer, Joshua (Josh) 22, 187
URS New Zealand 9, 67, 78, 80, 81–82, 87

Valley Longwall International 9, 101, 127, 156, 160, 187, 238, 240, 247, 248
Valli, Keith 15, 22, 183, 184, 186

van Rooyen, Pieter 9, 105, 153, 177
ventilation 40–41, 45, 72, 73, 74, 124–25, 127, 130–31, 140, 141, 174–75
   collapse of main shaft 75, 79–82, 83
   design 40–41, 72, 86–87, 144, 153, 154, 230
   ladder 75, 105, 112–14, 157–58, 203
   main fan 138, 143–46, 147, 160, 170, 235, 247
   main shaft 75–82, 112, 155, 158, 203–4, 225, 242, 247
   monitoring system 153, 154, 155, 193
   slimline shaft 105, 114, 196, 213, 218, 219, 232, 241
   surface fan (backup fan) 138, 144, 145, 146, 170, 193, 225
   see also Alimak raise
video footage 197, 203, 205, 208–9, 213–14, 214–15, 216, 217, 220, 232

Wallace, Jerry 142
Waratah Engineering 43, 86, 99, 100
   see also Juganaut loaders; *and under* continuous mining
Ward, Gordon 9, 28–29, 30, 31–32, 33–34, 39–40, 41, 42–43, 44, 45–46, 50, 51, 61, 68–69, 70–71, 83, 84, 93, 105, 108, 113, 114–15, 117, 119, 120, 129, 134–35, 148–49, 234, 245, 246, 247
Watts, Trevor 113, 195, 205, 218, 219, 220, 225, 226
Weir, Denise 38–39, 43, 93
White, Dave 221
White, Doug 9, 13, 15–16, 17, 122, 128–30, 131–33, 140, 143–44, 153–54, 156–57, 159, 161–62, 174–75, 178–80, 190–91, 192, 193, 196, 198,

205, 225, 228, 230, 243, 235
Whittall, Leanne 38, 94
Whittall, Peter 9, 246, 247, 248
　capital raising for Pike River Coal 38–41, 45, 46, 57–58, 63, 83, 89, 103, 114, 134–35, 150, 154, 160, 161–62, 191, 246
　charges against 238, 242, 247, 248
　early days of Pike River Coal 35, 37–38, 246
　mine operations and management 16, 69, 70, 87, 88–89, 92, 93–95, 98, 99, 104–5, 106–8, 111–12, 113, 123–24, 127–28, 129, 130, 132, 133, 137, 138, 140, 142, 144, 149, 150, 153, 159, 160–61, 163, 169, 170, 174, 176–77, 181, 246, 247
　post-tragedy 199–200, 202, 203–4, 206–7, 209, 210–11, 213–14, 215, 216, 218, 219, 220, 221–22, 223, 226, 228
　Royal Commission 234
Wilkinson, Kate 171, 238, 248
Williams, Carolyn 227
Williams, Sarah 78, 227
Wilson, Helen 227
Winter, Matt 180–81
Wishart, Brian 151
Wood, Denis 60, 61, 62, 246
Wylie, Stephen 148, 159, 161, 183–84